Computational Fluid Dynamics for
Wind Engineering

Computational Fluid Dynamics for Wind Engineering

R. Panneer Selvam, Ph.D., P.E.
University Professor
Department of Civil Engineering
University of Arkansas, Fayetteville, AR, USA

Registered Office(s)
111 River Street, Hoboken, NJ 07030, USA
The Atrium, Southern Gate, Chichester, West Sussex, PO19 8SQ, UK

Editorial Office
9600 Garsington Road, Oxford, OX4 2DQ, UK

For details of our global editorial offices, customer services, and more information about Wiley products visit us at www.wiley.com.

Wiley also publishes its books in a variety of electronic formats and by print-on-demand. Some content that appears in standard print versions of this book may not be available in other formats.

Library of Congress Cataloging-in-Publication Data
Name: Selvam, R. Panneer, author.
Title: Computational fluid dynamics for wind engineering / R. Panneer Selvam.
Description: Hoboken, NJ : Wiley-Blackwell, 2022. | Includes bibliographical references and index.
Identifiers: LCCN 2022012496 (print) | LCCN 2022012497 (ebook) | ISBN 9781119845058 (cloth) | ISBN 9781119845065 (adobe pdf) | ISBN 9781119845072 (epub)
Subjects: LCSH: Wind-pressure. | Computational fluid dynamics.
Classification: LCC TA654.5 .S4395 2022 (print) | LCC TA654.5 (ebook) | DDC 624.1/75–dc23/eng/20220624
LC record available at https://lccn.loc.gov/2022012496
LC ebook record available at https://lccn.loc.gov/2022012497

Cover Design: Wiley
Cover Image: © HelloRF Zcool/Shutterstock.com

Set in 9.5/12.5pt STIXTwoText by Straive, Pondicherry, India

SKY10035530_072922

Contents

Preface

My computational fluid dynamics (CFD) for wind engineering journey started around January 1983 at Texas Tech University (TTU) when myself and Dr. Kishor Mehta were brainstorming on new research areas on a Saturday morning and what I can consider for my PhD topic. Before that, I did not know anything about CFD and not much in fluid mechanics except taking a four-semester course work in my undergraduate program. Sincse I had reasonable background on numerical methods and its application to solid mechanics from my master's work in India and in the United States, I decided to apply those concepts to wind engineering applications. Especially, the tornado force on building fascinated me because only after I came to Lubbock, TX, I came to know about tornado and its devastation. In India where I grew up, I was exposed to hurricane-type wind extensively, and this may be another reason for me getting into wind engineering research area. At that time, I did not realize what I was getting into. Dr. Mehta did say I might not realize my dream even after 80 years old. However, Dr. James McDonald (my advisor) and Dr. Kishor Mehta did support my idea, and I started to apply numerical methods in fluid mechanics for tornado forces on buildings. I did not do any substantial work in my PhD work, but it did open the CFD application for wind engineering research area. My next vertical advancement happened when I visited Commonwealth Scientific and Industrial Research Organization (CSIRO), Australia, as a research scientist to work under Dr. John Holmes during the summer of 1990. He is a fun and nice person to work with, and I am glad he gave me an opportunity to work on CFD application to thunderstorm downdraft modeling. There I met Dr. David Peterson, and he taught me the implementation of the SIMPLE method of solving the Navier–Stokes (NS) equations and law of the wall boundary condition. There I used CFD to compute velocity in a thunderstorm downdraft and flow over 3D building using k-ε turbulence model. The paper (Selvam and Holmes 1992) becomes the beginning of thunderstorm downdraft study in wind engineering. From there on different challenges in CFD for wind engineering were resolved and now we are in a much better situation for application in wind engineering.

Dr. Allan Larsen in 1998, Dr. Partha Sarkar in 2010, and Dr. Prem Krishna in 2002, 2008, and 2017 requested me to write review papers on CFD for wind engineering. Those experiences gave me chances to reflect and advance myself for further developments. In the recent years, Dr. Arindam Chowdhury from Florida International University has become another motivator to expand my journey. Dr. Chowdhury and his student Dr. M. Moravej provided me wind tunnel data for the 1:6 scale TTU building, and he explained to me the partial turbulence simulation (PTS) method reported in Mooneghi et al. (2016) paper. This helped me to learn more about turbulence effects on building and challenges in wind tunnel modeling. He is a great person, and he opened my mind to learn more about inflow turbulence generation methods and energy cascade in turbulence. This is a concept many did not apply in turbulence modeling. If this concept were understood for practical application, the CFD application would have progressed much quicker. Murakami et al. (1987) used recycling method of considering turbulence in the flow using large eddy simulation (LES) for the first time in wind engineering. The recycling method has been used in many applications for several years after that. I also tried to implement it and reported my findings in Selvam (1997), and I thought at that time the turbulence energy has to be maintained as time goes on. From the numerical experiment, I found that after some time, most of the turbulence energy got lost in the computation. This could be due to the numerical diffusion as well as the energy cascade phenomenon. Because of my ignorance, I did not report the details in any of my publications. In recent years, I learnt that because of energy cascade and 3D modeling, the energy from lower frequency is transferred to higher frequency and also the waves get stretched and twisted.

In this work, random Fourier-based inflow turbulence generation method is used as inflow in Chapter 5 and the peak pressure on building is computed. The program developed for this case can be used for building aerodynamics study without inflow and with inflow. This helps the student to learn the power of CFD to some extent. This tool also gives a chance for students to generate their own wind data and analyze them for wind spectrum. The other notable problem considered is the vortex shedding in 2D cylinders. This provides a pathway to understand the vortex-induced vibration (VIV) issues in thin structures and bridges. The program for that also is used for class instruction. The programs developed for this class can run on a personal computer, and this makes it easier for students to use. The outputs are written in a format suitable for tecplot visualization program. The open-source visualization programs like ParaView can be used, and it is not user-friendly. However, the data can be manipulated for other systems easily because the files are in ASCII format.

To perform CFD modeling for building and bridge aerodynamics, some understanding of the NS equation, properties of turbulence, turbulence modeling, introduction to finite difference method, and wind engineering is necessary, and they are introduced briefly in Chapters 1–4. At the end of each chapter, necessary homework problems by hand or computers are provided to have hands-on experience.

A brief review of CFD application in wind engineering is provided in Chapter 6. I do apologize to many researchers whose work I could not include in Chapter 6 due to lack of time and space. In Chapter 7, use of OpenFOAM for wind engineering is introduced.

This course material was developed in the summer of 2020 to teach in the fall semester. Before the Fall of 2020, I taught CFD class twice, which helped me to develop the course material more focused toward wind engineering. The material for the class was expanded as the courses were taught. I had few fresh graduate students like Ms. Kaley Collins, Mr. Caleb Chestnut, and Mr. Gerardo Aguilar who gave a lot of support to teach this class in addition to my graduate students (Mr. Sumit Verma and Ms. Zayuris Atencio). Because of them, I got Mr. Andrew Deschenes, Mr. Wesley Keys, and Mr. Yancy Schrader in my class as students. The participation of all of them really improved the course material.

Even though the course material is more toward wind engineering application, if someone wants to write their own program, numerical algorithms are provided and several programs are listed for their own development.

The course was taught in the Fall of 2020 with our own CFD research code and tecplot up to Chapter 5. The students ran the programs on personal computer, and that made it easier for students. The visualization program tecplot is user-friendly, but it is a commercial program. If someone wants to teach the Chapter 5 material using open-source CFD program OpenFOAM and open-source visualization program ParaView, they can do so by using the material in Chapter 7. The major challenge may be to adopt an inflow turbulence generator available from OpenFOAM.

Since no other textbook on computational wind engineering is available at this time, I developed a teaching philosophy after several months of reflection. If you have any comments for improvement after going over the material, please email it to me. This means a lot and I greatly appreciate. I do hope this material is useful for students, industry practitioners, and researchers. I would like to thank Dr. A. Chowdhury for going over the material and providing valuable comments for improvement. Finally, I like to acknowledge the financial support received from Airforce, Navy, NASA, NSF, FHWA, James T. Womble Professorship and the Department of Civil Engineering, University of Arkansas over the years to conduct many of the research work reported in this book.

References

Murakami, S., Mochida, A., and Hibi, K. (1987). Three-dimensional numerical simulation of airflow around a cubic model by means of large eddy simulation. *Journal of Wind Engineering and Industrial Aerodynamics* 25: 291–305.

Mooneghi, M.A., Irwin, P., and Chowdhury, A.G. (2016). Partial turbulence simulation method for predicting peak wind loads on small structures and building appurtenances. *Journal of Wind Engineering and Industrial Aerodynamics*. 157: 47–62.

Selvam, R.P. (1997). Computation of pressures on Texas Tech Building using large eddy simulation. *Journal of Wind Engineering and Industrial Aerodynamics* 67 & 68: 647–657.

Selvam, R.P. and Holmes, J.D. (1992). Numerical simulation of thunderstorm downdrafts. *Journal of Wind Engineering and Industrial Aerodynamics* 44: 2817–2825.

R. Panneer Selvam
University of Arkansas
September, 2021

1

Introduction

Fluid mechanics and heat transfer have extensive application. From aeronautical industry to automatic industry, it is applied to several areas. Some of the notable areas are:

1) Aeronautical industry – design of airplane to electronic devices
2) Automobile industry
3) Building and bridge aerodynamics (Selvam 2017)
4) Electronic cooling (Silk et al. 2008; Sarkar and Selvam 2009)
5) Environmental flow and heat transfer
6) Metrological flow and weather prediction
7) Hydraulic flow
8) Water treatment (Liu and Zhang 2019)
9) Wind energy

In all areas, computer modeling has been extensively used in the recent years, and this branch of computation is called computational fluid dynamics (CFD). CFD provides the detail of velocities, pressure, and temperature at every point at each time in the computational domain. This helps to create animation in time and provides the detail of the flow changes in time. To gather this much of information from experiment is very expensive. In certain situation like weather prediction, we cannot do any experiment and computer simulation is the only tool to predict the weather. The major challenge in CFD is to develop a reliable computer model for a particular application. If this is established for a particular application, it will be very useful in the design of the system.

The CFD is applied from single-phase flow to multiphase flow. In the multiphase flow, it can be liquid–vapor flow, solid–liquid flow, and solid–liquid–vapor flow. In these flows, chemical reactions can occur. Some of the challenging flows I encountered in the past 30 years are

Wind–bridge interaction: Here the bridge moves due to wind and hence beyond certain velocities the bridge can flutter as reported in Selvam et al. (2002). Below the critical velocity for flutter, the bridge will not have unlimited oscillations. The concept of moving grid has to be used in addition to regular CFD modeling. The Tacoma Narrow Bridge failed due to flutter for a velocity of 64 km/h (17.8 m/s) in 1940. The critical velocity for flutter for Great Belt East Bridge is 252 km/h (70 m/s) as reported in Selvam et al. (2002). The critical velocity

Computational Fluid Dynamics for Wind Engineering, First Edition. R. Panneer Selvam.
© 2022 John Wiley & Sons Ltd. Published 2022 by John Wiley & Sons Ltd.

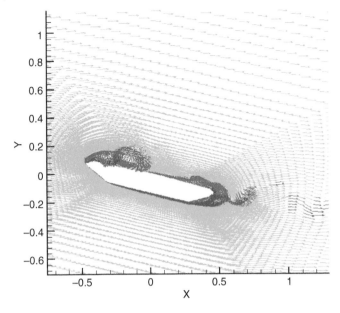

Figure 1.1   Flow around great Belt East Bridge during flutter condition.

depends upon the shape and structural properties of the bridge. The flow features during flutter condition are shown in Figure 1.1.

Heat transfer mechanism in spray cooling: Here, when a liquid droplet impacts a hot plate with a bubble growing in a thin liquid film; heat is removed due to complex interaction of droplet impact and vapor bubble. This high heat removal phenomena are explained in Selvam et al. (2006). For this, multiphase flow modeling of liquid and vapor is considered. In Figure 1.2 the liquid and vapor phases before the droplet impacts a vapor bubble in a liquid film are shown.

Tornado–building interaction: This study is reported in Selvam and Millett (2003, 2005). Here in a tornadic flow how a roof of a building is lifted up is explained using CFD. Figure 1.3 shows the velocity vector over the roof when a tornado-like vortex coincides with the center of a cubical building.

1.1   Brief Review of Steps in CFD Modeling

In the CFD modeling, the steps are very similar to well-established solid mechanics modeling. The major differences being most of the CFD applications are nonlinear and hence several iterations or time steps need to be performed.

Step 1: Grid Generation or Preprocessing: This may be the most time-consuming part if one has a complicated domain. If simple domain where in rectangular grid systems can be used, then the grid generation may be an easier task. Still one has to focus on the grid refinements in the boundary layer and in the regions of steep flow. Also, one has to make sure that grid spacing variation should not be high. The preferred ratio is 1.0–1.5. Very large ratios

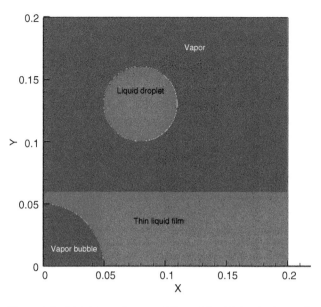

Figure 1.2 Multiphase flow modeling of liquid droplet impacting a vapor bubble in liquid film.

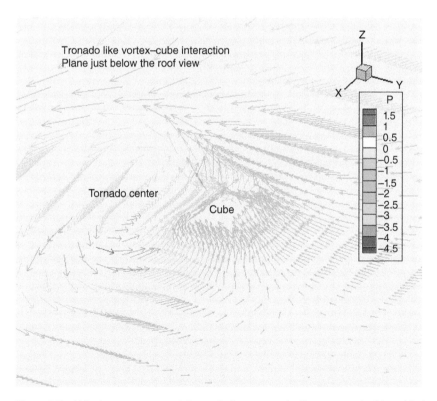

Figure 1.3 Velocity vectors around the roof when a tornado-like vortex coincides with the center of a cubical building.

like more than 5 or 10 are not preferable. For this step, extensive grid generation programs were developed in the recent years.

Step 2: Flow Solver: Once the grid is generated for a particular problem and the proper initial and boundary conditions are given for the problem, one can solve the Navier–Stokes (NS) equations. This is the most computer time-intensive step. For this several methods from direct to iterative procedures are developed to solve the Ax = b equations. To reduce computer time, high performance or parallel computing is also utilized. Sarkar and Selvam (2009) utilized parallel computing to reduce the computer time from 50 to 3 days for spray cooling applications. They also compared the performance of different iterative solvers in the parallel computing environment.

Step 3: Postprocessing: In this step, the output from flow solver is processed to mine valuable information. Here this can be done by regular x–y graphs, contours, vector plots, and the combination of all. If the data is written for several time steps for the whole region, one can make animation using software like TECPLOT, and flow features can be investigated. The flow visualization technique is very sophisticated and some time it is an art than science.

If it is a design, then one can change the parameters of the flow variable or computational domain and further computer runs can be made for further investigations.

Benefits of CFD:

1) Data available for all points in space and time.
2) Inexpensive comparing to experiment. Especially with the developments in computer speed and memory, CFD programs can run in a personal computer. The major hurdle is validating the CFD with experiment to have reliability.
3) Visualization and animation of data to understand the physical problem is easy to implement. This helps anyone to understand complex fluid phenomena.

1.2 CFD for Wind Engineering or Computational Wind Engineering

In wind engineering, the loads on building and bridges are obtained from wind tunnel (WT) measurements or field measurements. The field measurement is very expensive and only very limited field studies are conducted like Texas Tech University building. Currently, WT is the major tool used to investigate forces on buildings and to develop code regulations like ASCE 7-16. In recent years, CFD is emerging as an alternate tool. For more than 30 years, different researchers raised its capabilities and slowly it is becoming a reasonable tool to be used in wind engineering because of the availability of high-performance computers with large storage capacities. The work reported by Selvam (1992) took more than a day for one computer run. With the current computer capabilities, one can solve the same problem in few minutes. Hence, the speed increased may be more than 100 times in a single processor. With multiple processors, we can increase the speed at least 10 times. If the CFD model is well validated with experiments, then it becomes the most economical tool compared to experiments. The way finite element method (FEM) is used in solid mechanics area in the industry and research nowadays, the hope is someday CFD will be a tool in wind engineering. This book is a stepping stone to achieve the preceding objective.

To apply CFD in wind engineering, one needs to be familiar with the following topics:

1) Meteorology or atmospheric flow
2) Fluid mechanics
3) Turbulence
4) Random process or stochastic process
5) Numerical techniques like finite difference method (FDM) or FEM for fluid mechanics
6) Wind engineering
7) Visualization
8) Structural dynamics
9) Fluid–structure interaction
10) Water (wave-storm surge)–wind–structure interaction as in hurricane
11) Grid generation
12) Parallel computing

In the current work, we may not touch topics beyond point 7 in the preceding list because of lack of time. For the other topics, we will go into detail only what we use in our work. We use simple computational domain to reduce the difficulty of making proper grid. One can see the grid generation complexity in the wind–bridge interaction study, as shown in Figure 1.1. In the industry for complex 3D problem, one or two engineers may be spending two or three months to make a proper grid. Even in wind engineering, we only work on straight wind. We will not discuss much about the other types of winds (tornado and thunderstorm downdraft) due to lack of time. From 1960 onward, field observations and WT testing have been used to find pressures on building. Because CFD takes lots of computer time and memory, only recent years CFD application in wind engineering has emerged with more reliability.

In hurricane-type sever wind, in addition to wind effects on structures, water surge and wave effect produce enormous damage. This leads to multiphase flow (water and air) effect on buildings. Future application may involve water–wind effect on structures.

References

Liu, X. and Zhang, J. (2019). *Computational Fluid Dynamics: Applications in Water, Wastewater, and Storm Water Treatment*. ASCE Publication.

Sarkar, S. and Selvam, R.P. (2009). Direct numerical simulation of heat transfer in spray cooling through 3D multiphase flow modeling using parallel computing. *Journal of Heat Transfer* 131: 121007-1–121007-8.

Selvam, R.P. (1992). Computation of pressures on Texas Tech Building. *Journal of Wind Engineering and Industrial Aerodynamics* 43: 1619–1627.

Selvam, R.P. (2017). CFD as a tool for assessing wind loading. *The Bridge and Structural Engineer* 47 (4): 1–8. [Review paper-available as opensource].

Selvam, R.P. and Millett, P.C. (2003). Computer modeling of tornado forces on buildings. *Wind & Structures* 6: 209–220.

Selvam, R.P. and Millett, P.C. (2005). Large eddy simulation of the tornado-structure interaction to determine structural loadings. *Wind & Structures* 8: 49–60.

Selvam, R.P., Govindaswamy, S., and Bosch, H. (2002). Aeroelastic analysis of bridges using FEM and moving grids. *Wind & Structures* 5: 257–266.

Selvam, R.P., Lin, L., and Ponnappan, R. (2006). Direct simulation of spray cooling: effect of vapor bubble growth and liquid droplet impact on heat transfer. *International Journal of Heat and Mass Transfer.* 49: 4265–4278.

Silk, E.A., Golliher, E.L., and Selvam, R.P. (2008). Spray cooling heat transfer: technology overview and assessment of future challenges for micro-gravity application. *Energy Conversion and Management* 49: 453–468.

2

Introduction to Fluid Mechanics

Mathematical and Numerical Modeling

Any physical problem can be modeled by algebraic equations, differential equations (DEs), and partial differential equations (PDEs). This formulation of the physical problem into mathematical equation is called mathematical modeling. Using algebraic equations for mathematical modeling, the solution is achieved by simple one equation as in Pythagoras theorem or by simultaneous equations. The one degree of freedom structural dynamics equation is a DE problem, and two-dimensional solid mechanics problems discussed in theory of elasticity is a PDE example. For DE and PDE problems, there are analytical solutions in regular regions (square or rectangle) for linear problems. For example, the beam problem in structural analysis is a fourth-order DE. The simply supported beam with uniformly distributed load has closed-form solution. For complicated load or change in cross section, one may sought to numerical methods like moment area method or virtual work method. In the same way for 2D and 3D problems, numerical techniques like finite difference method (FDM) or finite element method (FEM) are used to reduce the DE and PDE to algebraic equations. There are problems such as quadratic equation is a nonlinear equation and special methods have to be sought to find solution. All of these issues are applicable for fluid mechanics problems, and we will discuss the mathematical modeling part in this chapter and the numerical technique part in the next chapter.

2.1 Navier–Stokes Equations

The Navier–Stokes (NS) equations are the basic governing equations for the fluid mechanics problems. Solving the equation provides the solution to all kinds of problems, and some of the practical problems are introduced in the first chapter. For simple problems, analytical approaches are used and that is what you would have used in your undergraduate work. It is very difficult to solve complex problems using analytical methods, and hence numerical technique is preferred. This text will attempt to solve the NS equations using numerical technique like control volume or finite difference procedure.

For the details of the derivation of the NS equations using basic scientific laws, one can refer to Anderson (1995), Versteeg and Malalassekera (2007), and Cengel and Cimbala (2006). Due to space limitations, only the complete compressible flow equation is reported here.

Computational Fluid Dynamics for Wind Engineering, First Edition. R. Panneer Selvam.
© 2022 John Wiley & Sons Ltd. Published 2022 by John Wiley & Sons Ltd.

To apply the compressible flow equations to practical problems, one needs to have good exposure not only in fluid mechanics but also in heat transfer. For basic understanding of heat transfer issues, one can refer to Incropera and DeWitt (1996). For the most of the incompressible flows, basic fluid mechanics exposure may be sufficient. The equations are reported in conservative and nonconservative form. Conservation form is preferred in the control volume procedures. Nonconservative form is used in the FEMs. For further discussion on advantages and disadvantages in using the aforementioned forms, one can refer to Anderson (1995) and other works.

2.2 Governing Equations for Compressible Newtonian Flow

$$\partial\Phi/\partial t + \partial(u\Phi)/\partial x + \partial(v\Phi)/\partial y + \partial(w\Phi)/\partial z - \partial(\Gamma\partial\Phi/\partial x)/\partial x$$

$$- \partial(\Gamma\partial\Phi/\partial y)/\partial y - \partial(\Gamma\partial\Phi/\partial z)/\partial z - S_\Phi \tag{2.1}$$

Time + (convection)	– (diffusion) Variable Φ	– source Γ	Source S_Φ	
Mass or continuity	ρ	0	0	(2.2)
Momentum	ρU	μ	$-\partial p/\partial x + S_x$	(2.3)
	ρV	μ	$-\partial p/\partial y + S_y$	(2.4)
	ρW	μ	$-\partial p/\partial z + S_z$	(2.5)
Internal energy	$\rho C_v T$	k	$-p\,\mathrm{div}\mathbf{u} + \varphi + Se$	(2.6)
Equation of state for perfect gas: $p = \rho RT$				(2.7)

Here:

$$S_x = -\partial(\mu\partial U/\partial x)/\partial x + \partial(\mu\,\partial V/\partial x)/\partial y + \partial(\mu\partial W/\partial x)/\partial z + \partial(\lambda\mathrm{div}\mathbf{u})/\partial x + f_x$$

where $\lambda = -(2/3)\mu$

Similarly, S_y and S_z can be developed

Dissipation function φ due to viscous stress $= \mu\{2[(\partial U/\partial x)^2 + (\partial V/\partial y)^2 + (\partial W/\partial z)^2]$

$$+ (\partial U/\partial y + \partial V/\partial x)^2 + (\partial U/\partial z + \partial W/\partial x)^2$$

$$+ (\partial V/\partial z + \partial W/\partial y)^2\} + \lambda(\mathrm{div}\mathbf{U})^2$$

Here, Se is the source term, $\mathbf{U}$ is the velocity in the vector notation, and speed of sound $c = \sqrt{(\gamma RT)}$.

When the flow is incompressible, the density ρ is constant and the governing equations for velocity and pressure are:

Continuity: div $\mathbf{U} = \nabla.\mathbf{U} = \partial U/\partial x + \partial V/\partial y + \partial W/\partial z = 0$

Momentum: $\rho[\partial\mathbf{U}/\partial t + \mathbf{U}.\nabla\mathbf{U}] + \nabla p - \mu\nabla^2\mathbf{U} = 0$

Momentum in the expanded form for 2D or in the x and y directions:

$$\partial U/\partial t + U\partial U/\partial x + V\partial U/\partial y + \partial p/\partial x - \partial(\nu\partial U/\partial x)/\partial x - \partial(\nu\partial U/\partial y)/\partial y = 0$$

$$\partial V/\partial t + U\partial V/\partial x + V\partial V/\partial y + \partial p/\partial y - \partial(\nu\partial V/\partial x)/\partial x - \partial(\nu\partial V/\partial y)/\partial y = 0$$

Comments and Observations from Fluid Mechanics Point of View:

1) There are totally six variables (ρ, U, V, W, T, and P) and six equations (2.2–2.7) are available to solve for each variable.

2) For **incompressible flow**, ρ and μ are constants, and hence div **U** = 0. The number of unknowns are five (U, V, W, T, and P). The five equations (2.2–2.6) without the equation of state will be used to solve the five variables. Also note that the energy equation does not influence the continuity and momentum equations. On the other hand, the energy equation is influenced by the velocities through viscous dissipation term. Hence many times if thermal issues are not involved, only continuity and momentum equations are considered for incompressible flow problems. If temperature issues need to be considered in the incompressible flow problems, knowing the velocity the energy equation can be solved independently.

3) **Compressible and incompressible flow regime and Mach number**: Mach number Ma is defined as the ratio of reference velocity to speed of sound (Ma = U/c). This term will be used to classify the flow either compressible or incompressible. Generally, if Ma > 0.3, then compressible flow equation has to be used. The reasoning to consider compressible effect in a flow for Ma > 0.3 is discussed in basic fluid mechanics in Fox and McDonald (1978). For air, when the velocity is around 100 m/s, then Ma = 100/346 = 0.29, and the density is affected by only 5% and hence incompressible flow assumption is reasonable.

4) Any PDE is not complete without proper **initial and boundary conditions (BCs)**. We will discuss them in more detail when we discuss individual problems. On solid walls, we use **no-slip BC**, i.e. when the wall is not moving, the velocities are zero for viscous flows and normal velocity is zero or **slip BC** for inviscid flow. The other BCs have to be developed from the governing equations.

5) Equation 2.1 is also classified into four parts. The first term is due to time effect, the second term is due to **convection** effect, the third term is due to **diffusion** effect, and the final term is called source term. The time and diffusion part are underlined, and the convection and source terms are italicized.

6) In computer modeling or numerical approximation of the PDE of Eq. 2.1, the major challenge is to approximate the convection term into algebraic equations. The methods in use introduce more numerical error than one expects unless finer grids are used. The rest of them are not as challenging. We will discuss further in detail in the upcoming chapters.

7) The aforementioned equations are highly **nonlinear** because of the convection term. If the flow is very slow and the convection terms are negligible or zero, the equation may become linear.

8) When $\mu = k = 0$, i.e. the viscosity and thermal conductive effects are neglected, then the equations are called the **Euler equations**. That is the only equation has time and convective effect and no diffusive and dissipative effect. When $\mu = 0$, the flow is called **inviscid** flow. Most of the undergraduate fluid mechanics works deal with inviscid flow. For many inviscid flows, simple solutions are derived with further assumptions.

The **Bernoulli equation** is derived assuming that the flow is inviscid, incompressible, and irrotational. In reality, the flow is viscous and many times turbulent. Only for simple one-dimensional flow like flow in a channel, closed-form solutions are derived including the viscous effect.

9) For the most of the high-speed flows, the viscous effect is only very close to the solid body. This is because on the stationary solid body the velocity is zero, and it is also called no-slip BC. Away from the body, the viscous effect is very less and hence it can be ignored. This is where the Euler equation may be applicable. But close to the body where the velocity goes from zero to maximum, the viscous effect is considerable. This region is also called **boundary layer** as shown in Figure 2.1. This can vary from few millimeters to meters depending on the type of problem. The grid resolution to capture the boundary layer is challenging task. At this time, only for smaller Reynolds number (Re < 3900? for flow over circular cylinder) this is possible. For high Re, one may need to use turbulence models to improve the accuracy with possible fine grid resolution. In Figure 2.1, if grid points are not taken to **capture the boundary layer,** then the results will not be realistic. In this work, the finest spacing close to the bridge is 0.00026 B where B is the width of the bridge. Also, once proper grid resolution is considered in the radial direction to produce necessary vortices, if proper grid resolution is not available in the tangential direction then vortices are not **transported properly,** and this is illustrated by comparing Figure 2.1a and b. In Figure 2.1a, 170 nodes are considered around the bridge and in Figure 2.1b 800 nodes are considered. One can see the flow feature differences, and accordingly the computed force coefficients differed. For further details, one can refer to Selvam (2010).

10) The aforementioned equation is viewed at this time mathematically. We should also look into physically as well as fluid mechanics application form. When we look for application, various approximations to the particular situation are applied to the aforementioned equations and hence various names.

2.3 Definition of Convection and Diffusion

Diffusion: Color powder dropped in a still water or a lake spread around close to a circle as shown in Figure 2.2. Here, there is no velocity in the flow, and hence U = 0 and the Reynolds number Re = $\rho UL/\mu$ = 0. The color diffuses in all directions equally for the case of Re = 0. Almost entirely by diffusion, the color is spread around. The error in numerically approximating the diffusion equation is less, and traditional central difference approximations are fine.

Convection: Color powder dropped in a running stream. Here, the color spreads downstream but not upstream as shown in Figure 2.2. The cases Re = ∞ is the straight-line path from the source, and there is no diffusion. For the case Re > 0 &<∞, the color spreads from the source as shown. There is more spread of color as one goes downstream due to diffusion in the flow. The effect of convection is seen predominantly than diffusion. The error in numerically approximating the convection equation is high because of discontinuity

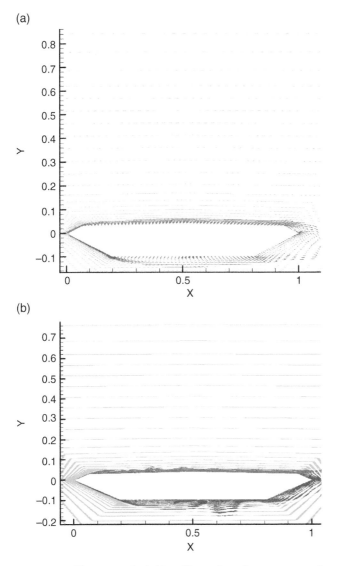

Figure 2.1 Flow around a bridge: Illustration of vortex generation and transport: (a) 170 nodes around the bridge and (b) 800 nodes around the bridge cross section.

introduced at the point where the color is introduced. That is there is no color in the upstream, and hence central difference approximations introduce more errors.

For *radiation*, there is no need of a medium to transport as in receiving the heat from sun. Also, this phenomenon is not directly considered in the NS equations. To have a good understanding of radiation, one can refer to Incropera and DeWitt (1996). The radiation effect is very important when the temperature difference between two bodies is very high, i.e. more than 200–300 °C.

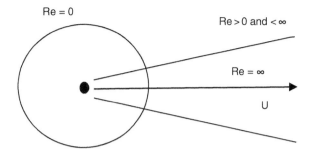

Re = 0

Re > 0 and < ∞

Re = ∞

U

Figure 2.2 Illustration of diffusion and convection.

2.4 Derivation of Bernoulli Equations

Let us assume the flow is incompressible as a start and the governing momentum equation in the vector form:

$$\rho[\partial \mathbf{U}/\partial t + \mathbf{U}\nabla\mathbf{U}] + \nabla p - \mu \nabla^2 \mathbf{U} - F = 0$$

$$\rho[\partial \mathbf{U}/\partial t + \nabla(\mathbf{U}^2/2) - \mathbf{U} \times (\nabla \times \mathbf{U})] + \nabla p - \mu \nabla^2 \mathbf{U} - F = 0$$

Here $\nabla \times \mathbf{U} = Wz$ is vorticity

If $\partial \mathbf{U}/\partial t = \mu = Wz = 0$, i.e. **steady, inviscid, and irrotational,** then the aforementioned equation reduces to:

$$\nabla(\rho \mathbf{U}^2/2 + P) = 0.0$$

Integrating we get $\rho \mathbf{U}^2/2 + p = $ constant.

So, this is valid only if the fluid is incompressible, steady, inviscid, and irrotational. Hence, if we know the velocity, we can calculate the pressure at any point. Most of the discussions in the first fluid mechanics undergraduate course are based on steady and inviscid flow.

2.5 Velocity Computation in an Incompressible, Irrotational, Steady, and Inviscid Flow

Continuity: $\nabla.\mathbf{U} = 0$, or $\partial U/\partial x + \partial V/\partial y = 0$

Vorticity: $\nabla \times \mathbf{U} = 0$ or $\partial V/\partial x - \partial U/\partial y = 0$

The two equations can be combined to solve the velocities.

Stream function form:

$$U = \partial\psi/\partial y, V = -\partial\psi/\partial x$$

The aforementioned relationship automatically satisfies the continuity, and the vorticity equation reduces to:

$$\partial^2\psi/\partial x^2 + \partial^2\psi/\partial y^2 = -Wz = 0$$

In the same way, one can formulate the following equation using potential function:

$$U = \partial\varphi/\partial x, V = \partial\varphi/\partial y$$

The aforementioned relationship automatically satisfies the vorticity equation, and the continuity equation becomes:

$$\partial^2\varphi/\partial x^2 + \partial^2\varphi/\partial y^2 = 0$$

For both the equations, one has to come up with a proper BC from the velocity. The common BCs for Poisson's equations from mathematical point of view are:

either ψ or $\partial\psi/\partial n$ specified or φ or $\partial\varphi/\partial n$ specified.

Using the aforementioned equations, the velocity around circular cylinder is solved analytically. Any basic fluid mechanics text like Cengel and Cimbala (2006) will give the details of the derivation as well as final solution. For any other shapes, one may need to use numerical techniques like FEM or control volume method (CVM). One should note that the aforementioned equation is **linear**, and hence the solution is arrived by solving $Ax = b$ equation one time. Knowing the velocities, the pressure is calculated from the Bernoulli equation. We will see the application later.

For the inviscid flow around circular cylinder due to constant velocity of U along x-axis, there is analytical solution available from any standard text book like Cengel and Cimbala (2006). The streamline ψ and the radial (Ur) and tangential (U_θ) velocities for a circular cylinder of radius a are given by:

$$\psi = U_r \sin\theta\left(1 - a^2/r^2\right), \; U_r = (1/r)\partial\psi/\partial\theta = U\cos\theta\left(1 - a^2/r^2\right) \text{ and}$$
$$U_\theta = -\partial\psi/\partial r = U\sin\theta\left(1 + a^2/r^2\right)$$

The streamline plot for the flow around the circular cylinder is shown in Figure 2.3 for inviscid flow as well as viscous flow for Re = 100. In the inviscid flow, there is no flow separation and the maximum velocity is on the circular cylinder. In addition, there is symmetry in the flow. Whereas, in the viscous flow, there is zero velocity on the cylinder and the velocity increases from zero on the wall to free stream velocity far away. In addition, there is flow separation behind the cylinder. This flow separation produces high pressure on the cylinder. In the same way in the building aerodynamics, this flow separation produces high pressures all around the building as shown in Figure 2.4.

2.6 Nondimensional NS Equations

In the computational fluid dynamics (CFD) modeling, mostly nondimensional equations are used. The nondimensional equations help to keep the change in variables close to 1, and hence there is more stability during computing. Especially, if the computation diverges slightly due to nonlinear nature, the maximum value does not reach the maximum allowed by the computer very quickly. Usually, the computers keep the range to be 10^{-20} or 10^{20}. Rather than changing several variables in the computation, one can change the

(a)

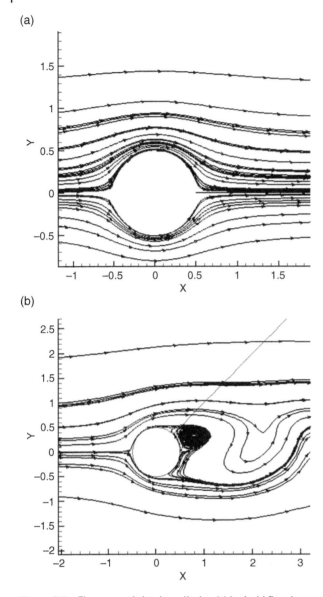

(b)

Figure 2.3 Flow around circular cylinder: (a) inviscid flow has no flow separation and (b) viscous flow at Re = 100 has flow separation.

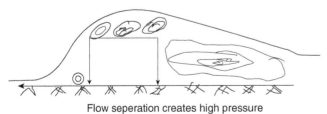

Flow seperation creates high pressure

Figure 2.4 Flow around a bluff body (building).

nondimensional parameters like Re to see the change in the flow. In the same way, in the analytical study also nondimensional equations are used. Here, we will derive the nondimensional equation for incompressible flow. For the other type of flow, one can derive it with the same approach.

Depending on the chosen reference values, one or more nondimensional parameters will arise in the NS equations. We will consider reference length as diameter of the cylinder D as in the flow over circular cylinder problem. The reference velocity is considered to be the far away velocity U_∞, and the density ρ and the viscosity μ of the fluid are kept constant. Here, the one-dimensional equation is taken for illustration, and finally we write the general equation.

Dimensional equation:

Continuity; $\partial U/\partial x = 0$

Momentum : $\rho[\partial U/\partial t + U\partial U/\partial x - \mu\,\partial^2 U/\partial x^2] + \partial p/\partial x = 0$

The momentum equation can be rewritten by dividing the ρ:

$\partial U/\partial t + U\partial U/\partial x - (\mu/\rho)\partial^2 U/\partial x^2 + \partial(p/\rho)/\partial x = 0$

Let us introduce nondimensional terms as $U^* = U/U_\infty$, $x^* = x/D$, $t^* = tU_\infty/D$ and $p^* = p/\rho U_\infty^2$

Then substitute $U = U^*\, U_\infty$, $x = x^* D$, $t = t^* D/U_\infty$ and $p^* = p/\rho U_\infty^2$ in the aforementioned continuity and momentum equation and factor the constants out:

$\partial U/\partial x = 0$ becomes $(U_\infty/D)[\partial U^*/\partial x^*] = 0$ and finally reduces to $\partial U^*/\partial x^* = 0$

$\partial U/\partial t + U\partial U/\partial x - (\mu/\rho)\partial^2 U/\partial x^2 + \partial(p/\rho)/\partial x = 0$

Taking common values for all the terms except diffusion, the equation becomes:

$(U_\infty^2/D)[\partial U^*/\partial t^* + U^*\partial U^*/\partial x^* + \partial p^*/\partial x^*] - [U_\infty\mu/\rho D^2]\partial^2 U^*/\partial x^{*2} = 0$

In the aforementioned equation, the coefficient for diffusion term has D^2 in the denominator because $\partial^2 U/\partial x^2$ is $\partial^2 U/(\partial x\partial x)$, and hence D^2 comes out in the denominator.

Dividing U_∞^2/D for all the terms, the equation becomes

$\partial U^*/\partial t^* + U^*\partial U^*/\partial x^* + \partial p^*/\partial x^* - [\mu/(\rho D U_\infty)]\partial^2 U^*/\partial x^{*2} = 0$

Defining Re= $(\rho D U_\infty)/\mu$, the equation becomes

$\partial U^*/\partial t^* + U^*\partial U^*/\partial x^* + \partial p^*/\partial x^* - [1/\,\mathrm{Re}]\partial^2 U^*/\partial x^{*2} = 0$

The nondimensional equation is written as follows without *:

Continuity; $\partial U/\partial x = 0$

Momentum : $\partial U/\partial t + U\partial U/\partial x - (1/\,\mathrm{Re})\partial^2 U/\partial x^2 + \partial p/\partial x = 0$

The general incompressible equation in the nondimensional form using vector notation:

Continuity : $\nabla.\mathbf{U} = 0$

Momentum : $\partial \mathbf{U}/\partial t + \mathbf{U}.\nabla\mathbf{U} + \nabla p - \nabla^2\mathbf{U}/\,\mathrm{Re} = 0$

Other text like Cengel and Cimbala (2006) uses different reference values, and they have more nondimensional numbers for the same equation. In our work, we will use this non-dimensional equation for computation.

2.7 Properties of Fluids

Knowing the properties of fluids helps to understand the influence of velocities, pressure, temperature, and density with one another. First, let us look at the properties of air.

2.7.1 Properties of Air

At normal temperature of T = 288 K or 15 °C

$$P = 101.3\,kPa, \rho = 1.225\,kg/m^3, \mu = 1.781 \times 10^{-6} kg/m.s$$

Temperature linearly varies with height. But, it decreases by 6.6 C/km. The density and pressure also decrease with the height.

2.7.2 Change in Velocity to Change in Energy

Change in velocity to change in pressure or density is less than 5% for M < 0.3.
Energy (in W.s/m^3) from velocity
$[\rho V^2 units\text{-}kg.\,m^2/(m^3 s^2)] = (kg.\,m/s^2).\,m/m^3 = N.\,m/m^3 = J/m^3 = W.\,s/m^3.$
Specific heat capacity Cp=1.007 kJ/(kg.k), ρ=1.16 kg/m^3 at 300 K.

U m/s	Energy in W.s/m^3	W.h/m^3
1	1	2.78×10^{-4}
10	100	2.78×10^{-2}
20	400	0.11
50	2 500	0.69
100	10 000	2.78

Overall change in velocity to change in energy is less compared to temperature or pressure or density.

2.7.3 Change in Temperature to Change in Energy

ΔT Temp.K   Energy $- \rho Cp \Delta T$ (J/m^3)

	W.s/m^3	W.h/m^3
1	1 000	0.278
5	5 000	1.39
10	10 000	2.78
20	20 000	5.56
50	50 000	13.9

2.8   Solution of Linear and Nonlinear Equations

To illustrate what is nonlinear equation, the flow in a nozzle is considered in the compressible range. First, the solution is considered as an incompressible flow and for that, Bernoulli equation for incompressible flow is used. Even though the incompressible Bernoulli equation is nonlinear, the solution is straightforward. Then the compressible flow is considered, and it is a nonlinear equation. One may need iteration to get converged solution. This exercise gives the taste of solving the nonlinear NS equations numerically.

Solution Using Bernoulli Equation for Incompressible Fluid (Linear Problem):

Continuity : $U_1 A_1 = U_2 A_2$

Energy : $P_1/\rho + U_1^2/2 = P_2/\rho + U_2^2/2$

A_i, U_i, and P_i are the area, velocity, and pressure, respectively, at point i.

Solution Using Compressible Fluid for Isentropic Flow (Nonlinear Problem):

Isentropic flow means: adiabatic and reversible flow or no energy loss or gain or constant entropy.

The following equations were taken from Anderson (1995) derivation of 1D compressible flow:

Continuity: $\rho_1 U_1 A_1 = \rho_2 U_2 A_2$ (E1)

Momentum: $p_1 A_1 + \rho_1 U_1^2 A_1 + (A_2 - A_1)p_{atm} = p_2 A_2 + \rho_2 U_2^2 A_2$ (E2)

Energy: $C_p T_1 + U_1^2/2 = Cp T_2 + U_2^2/2$ (E3)

Equationof State: $p = \rho RT$ (E4)

Speed of sound $C^2 = \gamma RT$, Mach number $M = U/c$

Here, $\gamma = 1.4$, and the gas constant R for air at room temperature of $T = 293$ K is 287 J/ (kgK). In the aforementioned four equations, we have four unknowns, and so we can solve for four unknowns if we know other four unknowns. Let us say we have inlet value and we calculate the outlet value. Since momentum and energy equations have velocities as a square, it becomes a nonlinear equation, and hence it is not that straightforward to solve. If we assume some reasonable velocity, then the remaining values can be solved comfortably if not it can diverge. We can understand some of the nonlinear issues by solving the equations in different form. As long as the values are very close to the final answer to start with, it will converge easily.

Example 2.1   Let us consider the flow through a converged nozzle as shown in Figure 2.5. In this example, we will compare the incompressible and compressible flow calculations. The column, which says Point 1, is the inlet conditions that are given. The known values are shown as (g). In the outlet (Point 2), only area is given and the velocity is assumed as a start for iteration. As a start, the incompressible U value got using Bernoulli equation is used. This calculation is reported in the last column.

Flow through a Nozzle

Properties	Point 1	Point 2-version1 (v1)	Point2-v2	P2-v3	Incompressible-p2
P kPa	84 (g)	82.98(s3-E4)	91.93	71.11	80.11
ρ kg/m^3	1 (g)	1.000 (s1-E1)	1.1	0.866	1.0
T in K	293 (g)	289.14(s2-E3)	290.7	286.19	
V m/s	**100 (g)**	**133.33 (assume)**	**121**	**154**	**133.33**
A m^2	0.012 (g)	0.009 (g)	0.009	0.009	
C m/s	343				
R J/(Kkg)	287				
Patm-kPa	101				
P-mom		91.5 (s4-E2)	91.5		
Error from momentum		81.86	147.6	−0.18	

Incompressible calculation details:

$$\text{From continuity}: U_2 = (\rho_1 U_1 A_1)/(\rho_2 A_2) = 1 \times 100 \times 0.012/(1 \times 0.009) = 133.33 \text{m/s}$$

$$\text{From Energy}: p_2 = p_1 + \rho(U_1^2 - U_2^2)/2 = 84000 + (100^2 - 133.33^2)/2 = 80111 \text{ Pa}$$

Compressible calculation details:

Assume U_2 =133.33 m/s from incompressible value. From that, calculate the other values in the following steps:

T_2 is from the energy equation, ρ_2 is from the continuity equation, and p_2 is from the equation of state. The p_2 also can be calculated from momentum. For convergence, the p_2 from the equation of state and momentum should be the same. In this case, there is an error of 9.3% for pressure p_2 with respect to correct value of $p_2 = 71.11$.

When U_2 is taken as 121.0 m/s, the $p_2 = 91.93$ kPa is almost close to 91.5 kPa. Then the density has an error of about 10% to incompressible assumption. The p_2 values calculated from the momentum equation did not change at all from version 1 to version 2. Hence, calculating the pressure from momentum may converge faster.

To explain the steps to calculate, these values symbol "s" is used in column 2. The continuity, momentum, energy, and equation state are identified as E1, E2, E3, and E4, respectively. If the following procedure is used, one can get improved solution:

Figure 2.5   Nozzle flow from left to right.

Point 1   Point 2

1) Assume velocity closer to incompressible value for Point 2. For example $U_2 = 133.33$ m/s.
2) Calculate the density ρ_2 using continuity as 1.0 kg/m^3 identified as (s1-E1).
3) Temperature T_2 from the energy as 289.14 K identified as (s2-E3).
4) Get p_2 from the equation of state as 82.98 kPa-identified as (s3-E4).
5) Get p_2 from the momentum equation as 91.5 kPa-identified as (s4-E2).
6) Check the error = RHS-LHS of the momentum equation. For $V_2 = 133.33$, the error is 81.86.
7) If step 6 is not close to 0, change the velocity again until you get a converged solution. For the aforementioned problem if the V_2 is increased from 133.33, the error decreases. When $U_2 = 154$, then the error reduces to $-.18$. This can be taken as the final answer.

So, one can conclude that because of nonlinearity in the equation, the convergence depends on initial values as well as how the error is reduced by the method. If velocity U_2 is taken as a starting value and calculate others based on this, the convergence may be better because the nonlinearity effect is reduced.

2.9 Laminar and Turbulent Flow

When the flow is smooth and adjacent layers of fluid slide one over the other, then the flow is called laminar. This happens when the flow Re (UL/ν) is less than the critical value.

Beyond the critical Re, the flow is turbulent. The flow properties vary with time at a point in a turbulent flow. Also, energy is dissipated far more in a turbulent flow than laminar flow. The time-varying velocity $U(t)$ is usually represented at a point as a sum of mean and time-varying component from the mean, that is $U(t) = U_{mean} + u'(t)$ as shown in Figure 2.6.

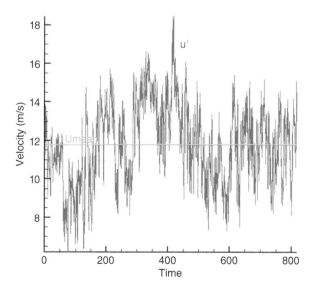

Figure 2.6 Velocity variation in time in a turbulent flow.

In turbulent flow, all the fluid properties (velocities, pressure, etc.) are characterized by mean values and statistical properties of time-varying part.

The flow changes from laminar to transition and then from transition to turbulent flow depending on the Re. The Re ranges from laminar to turbulent flow for certain common flows are as follows:

Transition to turbulent Re range	
Jet flow	Around 2000
Flow over a flat plate	10^3–10^6
Pipe flow	2×10^3 to 10^5
Flow over circular cylinder	200–400. As per Bloor (1964).
	Below Re = 200 is considered to be 2D flow

3D Modeling for Turbulent Flow

Turbulent flow is a 3D phenomenon. There are times 2D modeling is done, and this may be reasonable say for flow over circular cylinder at low Re and at some places to calculate the mean flow. Overall, for turbulent flow, the 3D modeling is necessary because of the following reasons:

1) The phenomena of energy cascade can happen only in 3D. This is an important energy transfer mechanism wherein the energy is transferred from low-frequency eddies to high-frequency eddies. In the 2D modeling of a wave transport, it was observed no loss in amplitude whereas in the 3D modeling, loss was observed when balanced tensor diffusivity (BTD) scheme reported in Smolarkiewicz (1968) was used.
2) To have energy cascade, more than one wave is needed in the inflow. If one wave is used at the inflow, the energy loss is not as substantial as reported in Selvam et al. (2020).
3) The flow features also change from 3D to 2D modeling because of the velocity in the third direction. When the 2D modeling of flow over a building is done by Woods (2019), large eddies formed behind the building because the energy could not be distributed. In the 3D modeling, that size of eddies are not observed.

2.10 Velocity Spectrum and Spectrum Considered by Different Turbulence Models

Any velocity data U(t) that is recorded in time as shown in Figure 2.6 can be represented in the frequency domain using Fourier analysis. The plot in Figure 2.6 is called time domain plot. We can have a plot of amplitude of the velocity for a particular frequency vs. frequency. Then, we say that the data is converted into frequency domain. This conversion is not only for time to frequency but also for space to frequency. Fourier first introduced the Fourier series in 1807 to solve the heat equation. Later for many years, it was the major tool to solve many science and engineering problems. Lord Rayleigh used it in his book Theory of Sound

(1877–1878) to solve many structural dynamics problems including the development of modal analysis.

Here, a brief introduction is provided. For more detail, one can refer to Paz and Leigh (2004). Fourier analysis is applied to many areas and is a powerful tool. Here, we will apply only to represent recorded velocity vs. time to velocity vs. frequency. A recorded velocity data in time can be represented by Fourier series as follows:

$U(t) = a_0 + \sum [a_n \cos(n\omega t) + b_n \sin(n\omega t)]$ for n = 1 to (np − 1)/2 where np is the number of points.

If dt is the time step, then total time $T = (np − 1)(dt)$, $\omega = 2\pi/T = 2\pi f$.

Here, f is the smallest frequency and T is the longest period. Normally, dt is taken to be a constant time step. The maximum frequency f_{max} for a given number of points Np will be $1/(2dt)$. This f_{max} is called Nyquist frequency. One should be aware that since the frequencies as well as coefficients a_n and b_n all increase by one number, if proper T for a given signal is not taken, the amplitude may be distributed around the critical frequencies. This is because the critical frequencies ($f_n = n/T$) may not be the same when T_1 different from T is taken. Then the new frequencies $f'_n = n/T_1$ will be different from $f_n = n/T$. This is a problem with Fourier series. For a good illustration, one should refer to Stull (1988).

The coefficients a_n and b_n can be calculated as follows:

$$a_0 = \left(\int U(t)dt\right)/T, \quad a_n = (2/T)\int U(t)\cos(n\omega t)dt \text{ and } b_n = (2/T)\int U(t)\sin(n\omega t)dt$$

$$(2.8)$$

Numerically, this can be calculated for constant dt as:

$$a_0 = \left(\sum U(t)dt\right)/T, \quad a_n = (2/T)\sum U(t)\cos(n\omega t)dt \text{ and } b_n = (2/T)\sum U(t)\sin(n\omega t)dt$$

$$(2.9)$$

Here, a_0 is the mean or average velocity U_{mean}. If the mean velocity is subtracted from U(t) ($u(t) = U(t) − U_{mean}$), then the amplitude can also be represented as:

$C_n = \sqrt{(a_n^2 + b_n^2)}$ and $u(t) = \sum C_n \sin(n\omega t + \varphi)$; here, φ is the phase angle.

The corresponding energy is C_n^2. The numerical way of calculating the Fourier coefficients using Equation (2.9) is called **discrete Fourier transform (DFT)**. Based on this, a program called **fs.f** is listed in the following section. This is a simple program but not a computationally efficient program. For efficient calculation, one should use **fast Fourier transform (FFT)**. A copy of the FFT program is listed in Cooley et al. (1969). In the FFT program, input data is given with a length of 2^n where n is an integer. The program **fs.f** does not calculate a_0 coefficients. This mean the data has to have mean value to be 0 or one can calculate the mean value separately.

Program **fs.f**

```
c.....PROG. FS.F, 8/14/2020
c      CALCULATE FS COEFFCIENTS
c.....PROG. SPD1.F, 9-24-17
c      CALCULATE SEPECTRAL DENSITY FUNCTION FOR EQUAL SPACING USING DFT
c.....period or frequencies using disc. fourier approach
```

```
c        freq3.f, calculates spectrum
         implicit real*8(a-h,o-z)
         open(5,file='fs-i.txt')
         open(2,file='fs-o.plt')
         phi=4.*atan(1.)
         phi2=2.*phi
         write(2,*)'freq,an,bn,amb'
         read(5,*)np,ttime,dt,nfreq
         fre=1./ttime
         do n=1,nfreq
         if(n.ne.1)read(5,*)
         fre1=fre*n
         an=0.0
         bn=0.0
         doi=1,np
         read(5,*)time,xa
         temp=n*phi2*time/ttime
         an=an+xa*cos(temp)
         bn=bn+xa*sin(temp)
         end do
         an=2.*an*dt/ttime
         bn=2.*bn*dt/ttime
         amb=sqrt(an*an+bn*bn)
         print *,fre1,an,bn,amb
         write(2,10)fre1,an,bn,amb
         rewind 5
         end do
10       format(4(1x,f10.4))
         stop
         end
```

Example 2.2   Rectangular pulse analytical approach.

$$f(x) = -1 \text{ in } -1 \le x \le 0 \, \& \, f(x) = 1 \text{ in } 0 \le x \le 1$$

Here $T = 2$, $\omega = 2\pi/T = \pi$

$a_n = 0$ and $b_n = 0$ if n even and $\mathbf{b_n = 4/(n\,\pi)}$ for n odd. Hence, $C_n = b_n$

$$f(x) = (4/\pi)\{ \sin(\pi x) + \sin(3\pi x)/3 + \sin(5\pi x)/5 + \cdots \}$$

The square pulse is drawn for n = 1, 3, 5, and 7 compared with exact function in Figure 2.7. This is the most difficult function to represent by Fourier series, and see how they converge to the exact value. The other interesting observation using Fourier series is how fast the amplitude approach zero. The amplitude b_n or C_n is plotted in Figure 2.8.

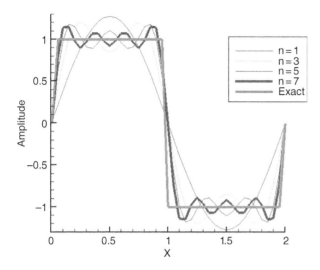

Figure 2.7 Square pulse plot for n = 1, 3, 5, 7, and exact.

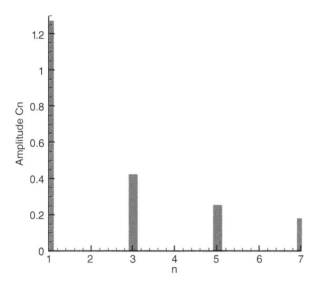

Figure 2.8 Plot C_n vs. n.

Example 2.3 Tent Function – Numerical Approach – DFT

Let us calculate the Fourier coefficients numerically using a tent function as a start. The tent function is shown in Figure 2.9. The wavelength L is 1.0, and the maximum amplitude is equal to 1.0. For DFT, we give only four points because numerically first and fourth points consider half of the other space. If you give five points, there will be some numerical error. Even though unequal spacing can be used for this calculation, equal spacing is more accurate for numerical integration.

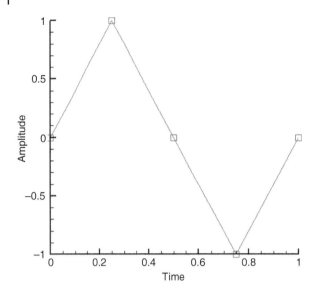

Figure 2.9 Tent function with five points.

The program **fs.f** has input file **fs-i.txt** and output file is **fs-o.plt**.
Input detail for the program **fs.f** in file **fs-i.txt**:

```
Line 1:read(5,*)np,ttime,dt,nfreq
      np           number of points not including the last point
      ttime        total time for np+1 points
      dt           time step
      nfreq        No. Of frequency to be considered usually np/2 or
                   less

Line 2 to np+1 lines: read each line time and f(t) for np lines.
do i=1,np
read(5,*)time,xa
end do

Sample data for np+1=5 points
4,1.,0.25,2              (Line 1 data)
0,0                      (Line 2 data)
0.25,1.0
0.5,0
0.75,-1
```

The corresponding coefficients using the **fs.f** program:
Output file: **fs-o.plt**.

```
Write nfreq lines:
write(2,10)fre1,an,bn,amb
```

```
fre1    frequency
an      value of an coefficient
bn      value of bn coefficient
amb     amplitude √(an²+bn²)
```

```
Sample output for np=4 data:
freq,an,bn,amb
     1.0000    -0.0000    1.0000    1.0000
     2.0000    -0.0000    0.0000    0.0000
```

So we get a1 = a2 = 0.0, b1 = 1.0, and b2 = 0.0.
Hence, f(t) = sin(2πt/L) = sin(2πt).
In this case, the answer seems exact but when np is increased, the error will show up.
Let us take np+1 = 9 points for the same wavelength L = 1.0.
The input data is:

```
8,1.,0.125,4
0,0
0.125,0.5
0.25,1.0
0.375,0.5
0.5,0
0.625,-0.5
0.75,-1
0.875,-0.5
```

The corresponding output:

```
freq,an,bn,amb
     1.0000    -0.0000    0.8536    0.8536
     2.0000    -0.0000    0.0000    0.0000
     3.0000     0.0000   -0.1464    0.1464
     4.0000    -0.0000   -0.0000    0.0000
```

Here f(t) = 0.8536sin(2πt) − 0.1464sin(6πt)
The plot of exact and DFT values are plotted in Figure 2.10.
For np = 8 points, the n = 1 function has some error and n = 3 has exact answer.
For np=16 points, the series comes to:

$$f(t) = 0.8211 \sin (2\pi t) - 0.1012 \sin (6\pi t) + 0.0452 \sin (10\pi t) - 0.0325 \sin (14\pi t)$$

The closed-form solution for this function:

$$f(t) = (4/\pi)\{ \sin (2\pi t) - \sin (6\pi t)/3^2 + \sin (10\pi t)/5^2 - \sin (14\pi t)/7^2 + \cdots.\}$$
$$f(t) = 1.273 \sin (2\pi t) - 0.1415 \sin (6\pi t) + 0.0509 \sin (10\pi t) - 0.0271 \sin (14\pi t)$$

Hence, there is a difference between closed-form solution and numerical or DFT solution.
Usefulness of Fourier series: The tool that is developed to represent any function by Fourier series can be used to draw the wind spectrum as it is discussed in the next section.

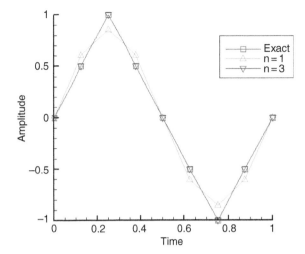

Figure 2.10 Tent function with nine points.

The importance of understanding wind spectrum will be very useful in selecting different turbulence models. Each turbulence model considers certain part of the wind spectrum, and the rest is modeled. This is also useful to understand and apply spectral method using Fourier series. The details of Fourier spectral method application to fluid mechanics are reported in Canuto (1988) and Peyret (2002).

Simple Application of Fourier Series

To understand the application of Fourier series for practical problems, let us consider two examples as follows:

Example 2.4 Simply supported beam with uniform loading w. Let us assume the beam has constant EI and the value is 1 for the ease of calculation. The governing equation for the BC for the beam problem is:

$$EId^4y/dx^4 = w \text{ for } 0 < x < L$$

At $x = 0$ and $x = L$, the BCs are $y = 0$ and $M = EId^2y/dx^2 = 0$.

The sine series $\sin(n\pi x/L)$ satisfies the required boundary conditions. Hence, we can have a solution of the form:

$$y = \sum y_n \sin(n\pi x/L)$$

The load w also can be represented by similar sine series: $w = \sum w_n \sin(n\pi x/L)$.

The values w_n can be calculated by multiplying both sides by $\sin(m\pi x/L)$ and integrating with limits 0 to L. The following integral relationships are used to get the final values:

$$\int \sin(m\pi x/L) \sin(n\pi x/L)dx \quad = 0 \text{ when } m \neq n$$

$$= L/2 \text{ when } m = n \text{ and } n \text{ is odd.}$$

$$= 0 \text{ when } m = n \text{ is even}$$

$$\int \sin(n\pi x/L)dx = 2L/n\pi \quad \text{for } n \text{ odd}$$

$$= 0 \quad \text{for } n \text{ even}$$

$$wn = (2/L) \int w(x) \sin(n\pi x/L)dx$$

Hence, LHS = $2wL/n\pi$ \qquad RHS = $w_n L/2$ \quad for n odd

$$w_n = 4w/n\pi$$

Substituting the y and w in terms of sine series into the governing equation, one get:

$$\text{LHS} = -(n\pi/L)^4 \sin(n\pi x/L) y_n/EI, \quad \text{RHS} = (4w/n\pi) \sin(n\pi x/L) \text{ for } n = \text{odd.}$$

Equating the proper terms, one get $y_n = 4wL^4/(EI(n\pi)^5) = 0.01307wL^4/EI(n)^5$.

Knowing y_n, one can get the displacement at any point in the beam. Let us check at $x = L/2$, we get $\sin(n\pi x/L) = 1$ or -1 for n odd. Hence, the final answer at $x = L/2$:

$$y(x = L/2) = 0.01307wL^4(1 - 1/3^5 + 1/5^5 - 1/7^5)/EI$$

$$y(x = L/2) = 0.01307wL^4(1 - 0.0041 + 0.00032 - 0.00006) = \mathbf{0.01302wL^4/EI}$$

The closed-form solution is $y = 5wL^4/(384EI) = \mathbf{0.013wL^4/EI}$.

Hence, the answer is almost the same with closed-form solution. If the loading is complicated and not easy to integrate by closed form, one can use numerical integration by DFT or FFT for sine series as follows:

$$w_n = (2/L) \int w(x) \sin(n\pi x/L)dx = (2dx/L) \sum (w_i \sin(n\pi x_i/L) \text{ for } n \text{ odd.}$$

Example 2.5 Here, we will consider one degree of freedom spring mass system for dynamic motion with sinusoidal loading. For details of the equation and solution, one can refer to Paz and Leigh (2004):

Governing equation: $md^2y/dt^2 + cdy/dt + ky = F_1 \sin(\omega't)$ with at $t = 0$, $y = 0$ and $dy/dt = 0$.

Here m is the mass, c is the damping coefficient, and k is the stiffness of the spring. The dynamic load factor (DLF = max. y_{dy}/y_{st}) is:

$$\text{DLF} = 1/\sqrt{\left[(1-r^2)^2 + (2r\xi)^2\right]}$$

Here:

max.y_{dy} = the maximum displacement due to dynamic load;

$y_{st} = F_1/k$ is the static displacement;

$r = \omega'/\omega$ where ω' is the applied angular frequency and ω is the angular frequency ($\sqrt{(k/m)}$) of the structure;

$\xi = c/c_{cr}$, where $c_{cr} = \sqrt{(2 \text{ km})}$.

For any external load like tent function with amplitude say 10k with a period say $T' = 0.5$ s, one can use Fourier series to represent the load as follows considering nine points:

$$F(t) = (10k) [0.8536 \sin (\omega't) - 0.1464 \sin (3\omega't)], \text{ where } \omega' = 2\pi/T' = 12.57 \text{ rad/s}$$

If the structural properties are $W = 100k$, $k = 200k/in$, and $\xi = 0.1$, then $T = 0.23$ s and $\omega = 27.79$ rad/s. Hence, $F_1 = 8.53k$ and $F_3 = 1.464$. The corresponding $y_{st1} = 0.043$ in and $y_{st3} = 0.007$ in can be easily calculated by dividing F_i by k.

So, the DLF for each frequency can be calculated as:

$$r_1 = \omega'/\omega = 12.57/27.79 = 0.452, \text{DLF}_1 = 1/\sqrt{\left[(1-r^2)^2 + (2r\xi)^2\right]} = 1.24$$

$$r_3 = 3r_1 = 1.356, \text{DLF}_3 = 1.135.$$

From DLF, we can say that ω'_1 has more effect than ω'_3. If one needs final maximum displacement, one can calculate y_{max1} and y_{max3} and sum them by absolute value or by RMS. In the same way for any loading, one can perform DFT to get the contribution of each frequency and perform the dynamic analysis. Before computers came, the only tool available was Fourier series.

Limitations of Fourier Series

1) The amplitude computed using DFT is the average in time. So, if the amplitude is changed for the same wavelength at different times, it will only show the average amplitude. This is because the DFT does not have memory of time. In wavelet analysis, this can be done more accurately. Look at the example in Selvam et al. (2020).
2) For a known wave with wavelength L, period T, and frequency f, if the exact length or time T of the wave is considered for DFT, the frequency will be calculated exactly. If the total time is taken little differently, then the amplitude will be distributed around the frequency for several frequencies. Also, the amplitude will be less than what is given. This is because if different ttime is taken, the fn=n/ttime will be different from the actual frequency. For further illustrations, one can refer to Stull (1988).

Wind Spectrum

To calculate the wind spectrum, we will use program **spd1.f** and compare with von Karman spectrum.

List the program **spd1.f**.

```
c.....PROG. SPD1.F, 9-24-17
c      CALCULATE SEPECTRAL DENSITY FUNCTION FOR EQUAL SPACING
       USING DFT
c.....period or frequencies using disc. Fourier approach
```

```
c       freq3.f, calculates spectrum
        implicit real*8(a-h,o-z)
        open(5,file='spd-i.txt')
        open(2,file='spd-o.plt')
        phi=4.*atan(1.)
        phi2=2.*phi
        read(5,*)np,ttime,dt,nfreq
        fre=1./ttime
        do n=1,nfreq
        if(n.ne.1)read(5,*)
        fre1=fre*n
        an=0.0
        bn=0.0
        doi=1,np
        read(5,*)time,xa
        temp=n*phi2*time/ttime
        an=an+xa*cos(temp)
        bn=bn+xa*sin(temp)
        end do
        an=2.*an*dt/ttime
        bn=2.*bn*dt/ttime
        amb=sqrt(an*an+bn*bn)
c.....compute spectrum as n*sx =fn*(dx/f1) where dx=(an*an
        +bn*bn)/2.
c.....df*sf=dx   where dx=(an*an+bn*bn)/2. Hence sf=dx/df-9/24/17
c        spe=fre1*(an*an+bn*bn)/(2.*fre)
        spe=(an*an+bn*bn)/(2.*fre)
c.....given spectra
        spe2=.779/(1.+50.*fre1)**1.667
c       print *,fre1,an,bn,amb
        print *,fre1,spe,amb
        write(2,*)fre1,spe,amb,spe2
        rewind 5
        end do
        stop
        end
```

The program **spd1.f** has input file **spd-i.txt** and output file **spd-o.plt**.
Detail of input data file spd-i.txt:

```
Line 1:read(5,*)np,ttime,dt,nfreq
      np         number of points not including the last point
      ttime      total time for np+1 points
      dt         time step
      nfreq      No. Of frequency to be considered usually np/2 or
                 less
```

```
Line 2 to np+1 lines:read each line time and f(t) for np lines.
doi=1,np
read(5,*)time,xa
end do
```

Detail of output data file **spd-o.plt**:
Output file: **spd-o.plt**.

```
Write nfreq
write(2,*)fre1,spe,amb,spe2
fre1    frequency
spe     spectral density value
amb     amplitude √(an²+bn²)
spe2    spectral density for a known function
```

In wind engineering, the dimensional frequency is represented by "n" and the nondimensional frequency is represented by $f = nH/U_{ref}$. Here, the reference length is H and reference velocity is U_{ref}. The plots can be an amplitude C_n vs. f or dimensional energy $(C_n^2/df = S(f))$ vs. f. If C_n or $S(f)$ is nondimensionalized, then the nondimensional value is C_n/U_{ref} and nS/U_{ref}^2.

To illustrate this, a field wind data is taken for 819.2 s. It is recorded at a frequency of 10 Hz or at 0.1 s. From this data, U_{ave} is calculated as 11.745 m/s. Then the data is converted to nondimensional value by considering reference height H = 4 m and the corresponding nondimensional dt=$0.1^*U_{ave}/H$ = 0.294 and ttime = 2408. The nondimensional velocities are plotted in Figure 2.11. The input details for the program **spd1.f** are:

$$NP = 8192 = 2^{13}, ttime = 2408, dt = 0.294 \text{ and } nfreq = 4000.$$

In Figure 2.11, the maximum frequency f is about 1.66 and the minimum frequency is around 0.0005. The minimum frequency here is $f_{min} = 1/ttime = 4.15 \times 10^{-4}$. The maximum frequency can be arrived from Nyquist frequency $f_{max} = 1/(2dt) = 1/(2^*0.294) = 1.7$. Normally, the wind spectrum can go up to f = 10. For that, recorded data should have high frequency. For a wind tunnel data received from Florida International University, the data is recorded for 2500 Hz. Then, f_{max} can go upto $nH/U_{ave} = 1250(4/8.6) = 581$. Also, the f_{min} from wind tunnel will be much lower in the range of 0.1. This plot gives you an idea which frequency has how much energy or amplitude from the wind. From this, one can also design the structures properly by avoiding severe resonance effect. The other observation is wind wavelength, and eddy size varies from L = 1/f = 0.6 to 2408 where L is the nondimensional wavelength. The actual wavelength can be calculated by multiplying reference length H as λ = 2.4 to 9632 m. Whereas in a WT, the wavelength may vary far less. Say H = 1.2 m and U = 21 m/s for a testing done by Mooneghi et al. (2016), the wavelength will be L = 0.1–10 and λ = 1.2–12 m.

Several spectral density functions are proposed from field observations. The one used in the program as a nondimensional spectrum for the longitudinal, lateral, and vertical turbulence at elevation z is from Kaimal et al. (1972) and is also mentioned in Simiu (2011):

$$nS_u(z,n)/u_*^2 = 200f/(1 + 50f)^{5/3}, \quad nS_v(z,n)/u_*^2 = 15f/(1 + 10f)^{5/3},$$
$$nS_w(z,n)/u_*^2 = 3.36f/\left(1 + 10f^{5/3}\right)$$

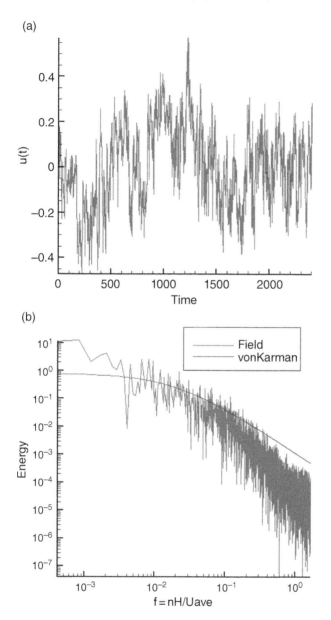

Figure 2.11 (a) Field velocity plot and (b) the corresponding spectrum.

Here, nondimensional frequency $f = nz/U(z)$ where n is the frequency. When plotting the wind spectrum, different work shows different nondimensional values for the y-axis. Mooneghi et al. (2016) used $nS_u/U(z)^2$.

In the plot, the U velocity spectrum is plotted. The equation used there is arrived as follows:

z = 4 m, U(4) = 11.475 m/s, and keeping z_o = 0.0375 m for open terrain, one get from U(4) = 11.475 = (u*/0.4) ln((4 + 0.0375)/0.0375). u* = 11.475 × 0.4/ ln[(4 + 0.0375)/0.0375] = 0.981.

$$S_u = 200u_*^2\, z\Big/\Big[U(4)(1 + 50f)^{5/3}\Big] = 200 \times 0.981^2 \times 4\Big/\Big[11.475(1 + 50f)^{5/3}\Big]$$
$$= 67.093/(1 + 50f)^{5/3}.$$

The aforementioned Kaimal spectrum is a function of f only. The spectrum reported in Aboshosha et al. (2015) has turbulence intensity (I) and turbulent length scale (Lu) also in the equation. The wind spectrum is a valuable information to understand the contribution of different turbulence models as well as in preparing input for inflow turbulence in the later chapters.

Method to Calculate the Amplitude from Wind Spectrum:

Variance or mean square for dimensional velocity $= \sum(a_n^2 + b_n^2)/2 = \int S_u(n)dn$. Hence $S_u(n)dn = (a_n^2 + b_n^2)/2$.
Dimensional $u'(t) = \sqrt{(2S_u(n)dn)}$ and nondimensional velocity $= u'(t)/U_{ave}$.
Correct the velocity spectrum plot accordingly.

Procedure using the Kaimal Spectrum:

1) Calculate the $S_u(n)$ from the given spectrum.
2) Multiply by 2dn and take the square root to get $u'(t)$.
3) Divide $u'(t)$ by U_{ave}. This is the nondimensional amplitude.

2.11 Turbulence Modeling

Turbulent flow is modeled by any one of the following ways:

1) Direct numerical simulation (DNS);
2) Large eddy simulation (LES);
3) Reynolds averaged NS (RANS) equation.

In the direct simulation, no turbulent viscosity term (μ_t) is added to the viscosity (μ) in the diffusion term of the NS equation. In the LES and RANS modeling, the μ_t term will be added to the viscosity term. This can vary every point in the computational domain. Extensive modeling work is done to calculate proper μ_t. The details of the reason behind it will be provided in the following discussions.

Direct numerical simulation (DNS): Here, the NS equation is solved without any turbulence models. Here, all the frequencies in the turbulence are resolved by the grid and are explained in Figure 2.12. The right-side unresolved frequency diagram is zero, and hence no need to use any turbulence models. The major challenge is to have necessary grid refinement whereever necessary to capture all the size of eddies that are developed. This limits the application to only certain flow with certain range of Re. For flow over circular cylinder, the maximum Re considered so far is 10 000 (Dong and Karniadakis 2005).

Large Eddy Simulation: The next level of sophisticated turbulence model is LES. Here, the time-dependent effect is captured reasonably with whatever the amount of grid size one can use. The eddy size that is greater than the grid size is captured, and the eddy size smaller than grid size is modeled using some sort of turbulence model as shown in Figure 2.12. Further discussions will be made in Chapter 5. The model is formulated from **space-filtered** equations.

Reynolds Averaged Equations: Here, the unsteady nature of the turbulence is modeled using **time-averaged** equations. The details of the time-averaged equations are given in Versteeg and Malalasekera (2007). As we discussed, the flow variable say U is split into time average mean and time-varying velocity from the mean, i.e. $U(t) = U_{mean} + u(t)$. To have a handle on the unsteady part $u(t)$, usually the statistical properties are monitored. That is the

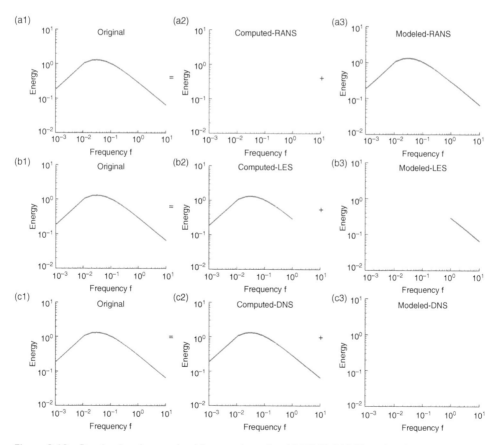

Figure 2.12 Resolved and unresolved frequencies using (a) RANS, (b) LES, and (c) DNS.

root mean square of u(t), i.e. $u_{rms} = \sqrt{}$(time average of u(t)^2) of the fluctuations, is considered. The kinetic energy k is a quantity often measured and is defined as:

$$k = \left(u_{rms}^2 + v_{rms}^2 + w_{rms}^2\right)/2$$

The turbulence intensity I is linked to the kinetic energy and a reference mean flow velocity U_{ref} as follows:

$$I = \sqrt{(2k/3)}/U_{ref}$$

Using the relation like U(t) = Umean + u(t), time average is taken for the NS equations. The time-averaged NS equation reduces to mean flow variables as unknowns. The NS equation is very similar to the original equation except that the variables are mean values and some extra terms are added in terms of the time-averaged products of the fluctuating velocities. These values occur from the time averaging of convection equation part. These terms are called **Reynolds stresses**, and these stresses are added to the terms due to viscous stresses in the computer modeling. The Reynolds stresses are:

$$Sxx = -\rho \text{ time average } \left(u^2\right), \quad Syy = -\rho \text{ time average } \left(v^2\right), \quad Szz = -\rho \text{ time average } \left(w^2\right)$$
$$Sxy = -\rho \text{ time average } (uv), \quad Syz = -\rho \text{ time average } (vw), \quad Sxz = -\rho \text{ time average } (uw)$$

More discussion on Reynolds stresses and integral length scales are found in Appendix B. Several turbulence models are formulated depending on the relationship developed for the Reynolds stresses in terms of the mean flow. Some of the Reynolds equation models are:

1) Zero-equation model – mixing length model
2) Two equation model – k-ε model
3) Reynolds stress equation model
4) Algebraic stress model

The two equation models are extensively used. They are based on the following relationships for the Reynolds stresses:

$$S_{ij} = -\rho \text{ time average } \left(u_i u_j\right) = \mu_t \left(\partial U_i/\partial x_j + \partial U_j/\partial x_i\right)$$

Here, μ_t is the turbulent viscosity. The assumption is that the turbulent stresses are related to the derivative of the mean flow. To close the problem, relationship has to be established for turbulent viscosity.

Zero-Equation Model: In the zero-equation model, a parameter called mixing length (lm) needs to be provided from experiment and experience, and it is called mixing length. The mixing length for different flows is given in Versteeg and Malalasekera (2007). Knowing the lm, the μt is calculated as follows:

$$\mu_t = lm^2 |(\partial U/\partial y|$$

In the recent days, it is not used very much in the computational problem.

k–ε Turbulence Model: In this model, two parameters, turbulent kinetic energy (k) and the rate of dissipation (ε), are used to calculate the μ_t:

$$\mu_t = C\mu\rho k^2/\varepsilon$$

Here, Cμ is a constant. The values of k and ε are calculated at each grid point by solving a transport equation similar to momentum equation reported in section 2.2. For details of the equations, one can refer to Versteeg and Malalasekera (2007) and Sorensen (1995). This model and its variants are extensively used in the engineering applications.

In the **Reynolds stress equation** model, six different transport equations are used to get the six Reynolds stresses. In the **Algebraic stress** model, approximate algebraic equations are used to calculate the stresses.

Note:

1) Using the Reynolds stress models, one can solve a problem as a steady state and get the turbulence statistics. This saves computer time.
2) Many papers are available wherein k-ε turbulence model is used for unsteady state problems like flow over circular cylinder at high Re. What I do not understand is when the NS equations are arrived from time average and one is solving for mean values, what is unsteady nature of the solution means. These things are not clear. In LES only, the eddies that cannot be resolved with the grid spacing are modeled using Reynolds stress relationships. So, time-dependent issues are captured.

LES Models

The subgrid-scale (SGS) eddies are modeled similar to Reynolds stress models and are called SGS Reynolds stress models. The SGS Reynolds stresses are modeled with various equations as reported in Piomelli (1999) and Geurts (2004). The simple one is the Smagorinsky model. This is similar to the eddy viscosity model reported before for time-averaged equations:

$$S_{ij} = -\rho \text{ space average} (u_i u_j) = \mu_t (\partial U_i/\partial x_j + \partial U_j/\partial x_i)$$

where $\mu_t = \rho(C_S\Delta)^2 |S|$ and $|S| = \sqrt{[(\partial U_i/\partial x_j + \partial U_j/\partial x_i)^2/2]}$

The term Δ is the filter width and C_s is a constant. The suggested C_s value is 0.1. The ratio of Δ/h can vary from 1 to 4 where h is the grid spacing in the FDM. For further discussions about the filter width, one can read the last chapter of Geurts (2004). The other sophisticated model used very much in engineering applications is dynamic model.

2.12   Law of the Wall

Define $u+ = U/u_\tau$, $y+ = \rho\, u_\tau y/\mu$, $u_\tau = \text{sqrt}(\tau/\rho)$.

Let us say the properties of air: $\rho = 1$ or $1.225\,\text{kg/m}^3$, $\mu=1.781\times 10^{-6}\,\text{kg/m.s}$.

Linear sublayer: $u+ = y+$ when y+ <5. Here $\tau = \mu U/y$, Hence $u_\tau^2 = \tau/\rho = \nu U/y$.

Log-law layer: $u+ = \ln(Ey+)/k$, where k = 0.4 and E = 9.8 for 30< y+ <500.

The validity and comparison of the two relations are reported in standard fluid mechanics text as well as in Versteeg and Malalasekera (2007).

Let us consider an example of boundary layer flow:

open terrain $z_0 = 0.035$ m, $k = 0.4$, $U(10\text{ m}) = 10$ m/s

$U(z) = (u_\tau/k) \ln[(z + z_0)/z_0].$

$U(10) = 10 = (u_\tau/0.4) \ln[(10 + 0.035)/0.035]; u_\tau = 0.707$

$U(0.01\text{ m}) = 0.44$ m/s

Note:

1) The aforementioned equation gives $U(0) = 0$, and the log-law layer does not give zero. Check!?
2) For nondimensional problem, $z_0 = 0.035/L_{ref}$.
3) For the tornado flow problem, Gorecki and Selvam (2014) have a non-dimensional grid spacing of 0.001 units with $L_{ref} = 10$ m, so the first grid spacing is 0.01 m.
4) Using $u_\tau = 0.707$ if I calculate $y+$, $y+ = u_\tau y/v = 0.707 \times 0.01/1.78 \times 10^{-6} = 3970$.

Procedure to Calculate y+:

1) Assume $u+ = y+$; calculate $u_\tau^2 = \tau/\rho = vU/y = 1.78 \times 10^{-6} \times 0.44/0.01 = 78.32 \times 10^{-6}$, $u_\tau = 8.85 \times 10^{-3}$, Hence, $y+ = u_\tau y/v = 8.85 \times 1 - 03 \times .01/(1.78 \times 10^{-6}) = 49.72 > 30$, and hence it is in the log-law layer.
2) Use the calculated $y+$ to calculate $u+$ using log-law $u+ = 15.47$. Use this to calculate $u_\tau = 0.028$ and then $y+$ (160) until it converges.

2.13 Boundary Layer Depth Estimation

Laminar flow : $d/L_{ref} = k/\text{sqrt}(\text{Re})$

Turbulent flow : $k/(\text{Re})**(1/7)$ or $k/(\text{Re})**(1/5)$

Here, k is a constant. In some of my papers, I proposed to use the first grid point spacing to be as in the laminar flow equation.

The aforementioned equations are mentioned for boundary layer depth on a flat plate in Cengel and Cimbala (2006).

2.14 Chapter Outcome

1) Familiar with incompressible NS equations.
2) Know about diffusion, convection, and unsteady nature of the NS equations.
3) Mathematical modeling of wind engineering problem by NS equations.

4) Exposed to difficulty in solving the nonlinear equations.
5) Fourier series for wind engineering – express a given wind data into Fourier series using DFT.
6) Wind spectrum from wind tunnel and field data.
7) Turbulence modeling. Reynolds stresses and turbulent length and timescale from Appendix B.
8) Boundary layer depth estimation for CFD calculations.

Problems

1   For similar input to the flow through nozzle problem, what is the output for $V = 50$ and 150 m/s instead of $V = 100$ m/s and compare it with Bernoulli equation.

2   If you try other methods iteratively, please report and we will discuss in the class. I try to solve ρ, V, T, and P sequentially with initial values on the inlet and it diverged.

3   Get the governing equations with proper BCs for Poiseuille flow and Couette flow. Derive from incompressible NS equation and state the assumption.

4   Calculate the Fourier coefficients for an isosceles triangle with $L = 1.0$ and amplitude $= 1.0$. Consider $h = L/4$ and $L/8$.

5   When two triangles from problem 4 with one inverted, we get tent function. If the same two triangles are apart say by 1 unit, then it will have $L = 3$. Perform DFT with the amplitude of the triangles as 1 unit as shown in Figure 2.13. Consider $h = L/6$ and $L/12$.

6   Find the dominant frequencies for the given time vs. u(t) data by DFT analysis:

t	0	0.125	0.25	0.375	0.5	0.625	0.75	0.875	1
u	0	1.207107	1	0.207107	0	-0.20711	-1	-1.20711	-4.9E-16

7   Calculate the wind spectrum using **spd1.f** for the given wind data C_n vs. f and S vs. f.

8   Calculate the y+ for different depth 0.001 and 0.005 m.

9   Need to know the details of different models.

10   Find the largest normal grid spacing close to the wall recommended for a problem with Re=100 and Re=1000. Here, use $dr = 1/[10 \sqrt{Re}]$.

(a)

(b)

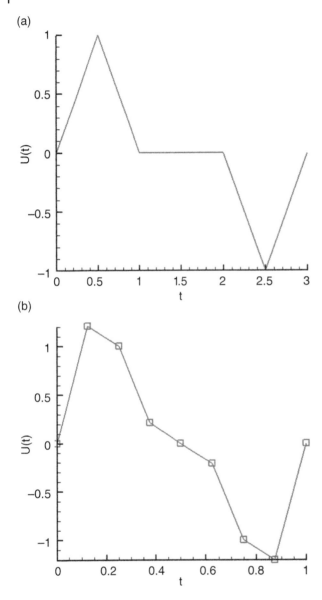

Figure 2.13 (a) Plot of a tent function with a gap for problem 5 and (b) plot for the data given in problem 6.

References

Aboshosha, H., Elshaer, A., Bitsuamlak, G.T., and El Damatty, A. (2015). Consistent inflow turbulence generator for LES evaluation of wind-induced responses for tall buildings. *Journal of Wind Engineering and Industrial Aerodynamics* 142: 198–216.

Anderson, J.D. (1995). *Computational Fluid Dynamics*. McGraw Hill.

Bloor, M. (1964). The transition to turbulence in the wake of a circular cylinder. *Journal of Fluid Mechanics*, 19(2), 290–304.

Canuto, C. (1988). *Spectral Methods in Fluid Dynamics*. New York: Springer-Verlag.

Cengel, Y.A. and Cimbala, J.M. (2006). *Fluid Mechanics: Fundamentals and Applications*. McGraw Hill.

Cooley, J.W., Lewis, P.A.W., and Welch, P.D. (1969). The fast Fourier transform and its applications. *IEEE Translations in Education* E-12: 27–34.

Dong, S. and Karniadakis, G.E. (2005). DNS of flow past a stationary and oscillating cylinder at Re = 10 000. *Journal of Fluids and Structures* 20: 519–531.

Fox, R.W. and McDonald, A.T. (1978). *Introduction to Fluid Mechanics*, 2ee. John Wiley &Sons.

Geurts, B. (2004). *Elements of Direct and Large-eddy Simulation*. Philadelphia, PA: R.T. Edwards.

Gorecki, P.M. and Selvam, R.P. (2014). Visualization of tornado-like vortex interacting with wide tornado-break wall. *Journal of Visualization* 18: 393–406.

Incropera, F.P. and DeWitt, D.P. (1996). *Fundamentals of Heat and Mass Transfer*, 4ee. New York: John Wiley & Sons.

Kaimal, J.C., Wyngaard, J.C., Izumi, Y., and Coté, O.R. (1972). Spectral characteristics of surface-layer turbulence. *Quartely Journal of the Royal.Meteorological Society* 98 (417): 563–589.

Mooneghi, M.A., Irwin, P., and Chowdhury, A.G. (2016). Partial turbulence simulation method for predicting peak wind loads on small structures and building appurtenances. *Journal of Wind Engineering and Industrial Aerodynamics* 157: 47–62.

Paz, M. and Leigh, W. (2004). *Structural Dynamics: Theory and Computation*, 5ee. Kluwer Academic Publishers.

Peyret, R. (2002). *Spectral Methods for Incompressible Viscous Flow*. Springer, Also appeared as: Applied mathematical sciences Vol. 148.

Piomelli, U. (1999). Large-eddy simulation: achievements and challenges. *Progress in Aerospace Sciences* 35: 335–362.

Selvam, R.P. (2010). Building and bridge aerodynamics using computational wind engineering. *Proceedings: International Workshop on Wind Engineering Research and Practice*, Chapel Hill, NC, USA (28–29 May).

Selvam, R.P., Chowdhury, A., Irwin, P., Mansouri, Z., and Moravej, M. (2020).CFD peak pressures on TTU building using continuity satisfied dominant waves (CSDW) method as inflow turbulence generator. Report, Department of Civil Engineering, University of Arkansas.

Simiu, E. (2011). *Design of Buildings for Wind: A Guide for ASCE 7-10 Standard users and Designers of Special Structures*. Hoboken, NJ: Wiley.

Smolarkiewicz, P.K. (1968). The multi-dimensional Crowley advection scheme. *Monthly Weather Review* 110: 1968–1983.

Sorensen, N.N. (1995). *General Purpose Flow Solver Applied to Flow Over Hills, RISO-R-827 (EN)*. Roskilde, Denmark: Riso National Laboratory.

Stull, R.B. (1988). *An Introduction to Boundary Layer Meteorology*. Springer.

Versteeg, H.K. and Malalasekera, W. (2007). *An introduction to Computational Fluid Dynamics*, 2ee. Prentice Hall.

Woods, J. (2019). *Turbulent Effects on Building Pressure using a Two-Dimensional Finite Element Program*. Civil Engineering Undergraduate Honors Theses Retrieved from https://scholarworks.uark.edu/cveguht/54.

3

Finite Difference Method

3.1 Introduction to Finite Difference Method

We saw in the previous chapter, the wind effects on structures are represented by the Navier–Stokes (NS) equations. This means a physical problem is modeled as a partial differential equations (PDE) with proper initial and boundary conditions (BCs) from the mathematical point of view. The solution of this complicated equation is achieved by numerical methods because there is no classical solution available for practical problems. There are several numerical methods available to approximate the PDE such as finite difference methods (FDM), finite element methods, and spectral methods. In this chapter, we will discuss FDM to approximate the NS equations as algebraic equations. Then one has to find a way to solve these simultaneous equations and a brief introduction to solver is also provided. In the process, the possible problem you will be exposed in this chapter are follows:

1) Potential flow for two-dimensional (2D) application. Flow over a square cylinder or flow in and out of a cavity. Introduction of Gauss–Siedel, SOR, and line iteration methods. Biringen and Chow (2011) have the similar illustration for flow in and out of a cavity. This illustrate the **steady-state solution using diffusion equation**.
2) Introduce 1D wave equation to explain **convection process, explicit and implicit method of solution**. The effect of each method on solution and error. Introduce Gauss–Siedel and line iteration.
3) Solution of NS equation. Overall discussions without numerical detail are provided.

Since approximating the complex NS equations as algebraic equations on a grid using FDM for hand calculation is very difficult, we will illustrate in this chapter the properties of the diffusion and convection part of the NS equation numerically. Even out of the two, only diffusion equation is illustrated by hand and the convection equation by computer program. The solution of the NS equation is illustrated intuitively.

Computational Fluid Dynamics for Wind Engineering, First Edition. R. Panneer Selvam.
© 2022 John Wiley & Sons Ltd. Published 2022 by John Wiley & Sons Ltd.

3.2 Example for 2D Potential Problem and Solution of Simultaneous Equations-Direct and Iterative Methods

Let us assume that the flow in a chamber to be inviscid and irrotational. Then the continuity equation and vorticity equation in 2D are the governing equations as we discussed in Chapter 2, and they are as follows:

Continuity equation : $\partial u/\partial x + \partial v/\partial y = 0$

Vorticity equation : $\partial v/\partial x - \partial u/\partial y = \omega_z$

The relationship between velocities and stream function is $u = \partial\psi/\partial y$ and $v = -\partial\psi/\partial x$
Substituting into continuity equation automatically satisfies the relation, and the vorticity equation $\partial v/\partial x - \partial u/\partial y = \omega_z$ becomes:

$$\partial^2\psi/\partial x^2 + \partial^2\psi/\partial y^2 = -\omega_z$$

For an irrotational flow, $\omega_z = 0$ and hence the equation becomes $\partial^2\psi/\partial x^2 + \partial^2\psi/\partial y^2 = 0$
The possible **BCs** for the preceding equation is either ψ specified or $\partial\psi/\partial n$ (tangential velocity) specified. Here n is the normal derivative.

The ψ specified boundary condition is also called Dirichlet BC or essential BC. In the case of structural mechanics, the equivalent BC is displacement specified. Similarly, specifying $\partial\psi/\partial n$ is also called Neuman BC or nonessential BC. In the structural mechanics, it is called force or stress specified.

The preceding problem can be solved by FDM or control volume method (CVM), finite element method (FEM), and spectral method. Here we will have brief introduction to FDM because it is easy to learn and implement in a rectangular grid system. The computational time will be less compared to other methods. The FEM can be applied to irregular region and takes more computer time. For spectral method, one can refer to Canuto (1988).

3.3 Finite Difference Method of Approximating the Partial Differential Equation

3.3.1 Introduction to Finite Difference Method

Using Taylor series one can derive the approximations to differential and PDE. To explain this, let us approximate the function ψ at point i as shown in Figure 3.1. We assume the

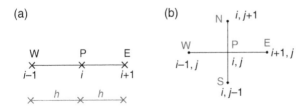

Figure 3.1 (a) 1D finite difference grid (b) 2D finite difference grid.

points are spaced at equal distance h (distance $x_{i+1} - x_i$ or $x_i - x_{i-1}$ is the same). In identifying these points instead of using i, i + 1 and i − 1 notation, one can also use center point i as P and i + 1th point as east point (E) and i − 1th point as west point (W), as shown in Figure 3.1. Then for three-dimensional (3D) one can use for (i,j,k) point as (P), (i + 1,j,k) point as east (E) point, (i − 1,j,k) point as west (W) point, (i,j + 1,k) point as north (N) point, (i,j − 1,k) point as south (S) point, (i,j,k + 1) point as top (T) point, and (i, j,k − 1) point as bottom (B) point. For programming ijk notation is useful and for general references, the other one may be useful.

Let us expand the ψ value at grid point i − 1 (ψ_{i-1}) and i + 1 (ψ_{i+1}) about point i (ψ_i) using Taylor series.

$$\psi_{i+1} = \psi_i + h(\partial\psi/\partial x)_i + (h^2/2)(\partial^2\psi/\partial x^2)_i + \text{higher order terms}$$

$$\psi_{i-1} = \psi_i - h(\partial\psi/\partial x)_i + (h^2/2)(\partial^2\psi/\partial x^2)_i + \text{higher order terms}$$

By subtracting one equation with another one, one can get:

$$(\partial\psi/\partial x)_i = (\psi_{i+1} - \psi_{i-1})/(2h) = (\psi_E - \psi_W)/(2h)$$

By adding the two equations one can get:

$$(\partial^2\psi/\partial x^2)_i = (\psi_{i+1} - 2\psi_i + \psi_{i-1})/h^2 = (\psi_E - 2\psi_P + \psi_W)/h^2$$

Algebraic expressions for higher order derivatives, one can refer to Crandall (1956). The derivatives can be represented nicely as a stencil shown below:

		ψ_{i-2}	ψ_{i-2}	ψ_i	ψ_{i+1}	ψ_{i+2}	
$(\partial\psi/\partial x)_i$	=			$[-1$	0	$1]/(2h)$	
$(\partial^2\psi/\partial x^2)_i$	=			$[1$	-2	$1]/h^2$	
$(\partial^3\psi/\partial x^3)_i$	=		$[-1$	2	0	-2	$1]/(2h^3)$
$(\partial^4\psi/\partial x^4)_i$	=		$[1$	-4	6	-4	$1]/h^4$

Using the preceding relationship, we can also express the second derivative in the y direction with subscripts j. Then for a grid in the x and y directions with equal spacing h, can be expressed with two subscripts (details can be seen in Figures 3.2 and 3.16) as follows:

$$(\partial^2\psi/\partial x^2)_{i,j} + (\partial^2\psi/\partial y^2)_{i,j} = \left(\psi_{i+1,j} - 2\psi_{i,j} + \psi_{i-1,j}\right)/h^2 + \left(\psi_{i,j+1} - 2\psi_{i,j} + \psi_{i,j-1}\right)/h^2$$

Simplifying:

$$(\partial^2\psi/\partial x^2)_{i,j} + (\partial^2\psi/\partial y^2)_{i,j} = \left(\psi_{i+1,j} + \psi_{i-1,j} + \psi_{i,j+1} + \psi_{i,j-1} - 4\psi_{i,j}\right)/h^2$$

$$= (\psi_E + \psi_W + \psi_N + \psi_S - 4\psi_P)/h^2$$

For unequal spacing and for more much accurate way of approximating the derivatives, one can use CVM. The detail of CVM method is well illustrated in Patankar (1980) and Ferziger and Peric (2002). We will use these algebraic expressions to approximate the partial derivatives at any point in a grid.

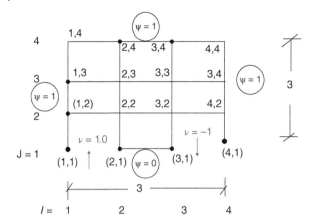

Figure 3.2 Symmetric flow in and out of a cavity for a 4 × 4 grid for hand calculation.

All the preceding FDM approximations are called central difference scheme, and they have second-order accuracy for $d\psi/dx$. At the end of a boundary, say the west point, we want to calculate the first derivative, then we can use forward difference:

$$(\partial\psi/\partial x)_W = (\psi_P - \psi_W)/h$$

This has first-order accuracy. Same way, we can use backward difference for the right-side boundary.

3.3.2 Physical Problem and Modeling

The physical problem is the flow that goes into a chamber one way and comes out another way, as shown in Figure 3.2. We will keep the inlet and outlet width to be of unit length. The total size of the chamber is a square with a length of 3. We will use a grid spacing of h = 1 for hand calculation illustration. This makes the grid to be 4 × 4. To model a physical problem and to illustrate the direct and iterative method of solution, we will consider a potential problem example by hand and as well as using computer program.

In the chamber, the flow is inviscid and irrotational, and the governing equation as discussed in the previous section is

$$\partial^2\psi/\partial x^2 + \partial^2\psi/\partial y^2 = 0$$

The corresponding algebraic approximation using neighboring values are

$$\left(\psi_{i+1,j} + \psi_{i-1,j} + \psi_{i,j+1} + \psi_{i,j-1} - 4\psi_{i,j}\right)/h^2 = 0$$

Hence, the preceding equation is rewritten as

$$4\psi_{i,j} - \left(\psi_{i+1,j} + \psi_{i-1,j} + \psi_{i,j+1} + \psi_{i,j-1}\right) = 0 \tag{3.1}$$

In the preceding representation, i and j notations are used to represent the 2D grid system.

Example 3.1 The flow in and out of the chamber is shown in Figure 3.2.

The BCs are derived from known conditions all around. Between point (1,1) and point (2,1), we say the flow is going upward.

Hence, $v = -\partial\psi/\partial x = -(\psi_{2,1} - \psi_{1,1})/h = 1.0$.

Keeping $\psi_{2,1} = 0$, $\psi_{1,1} = 1$ and $h = 1$, we will get $v = 1.0$. This means the bottom boundary between (2,1) and (3,1) is kept $\psi = 0$, and all other points we keep $\psi = 1$. This provides automatically $v = -1$ between points (3,1) and (4,1). Knowing the boundary values, we can solve for the interior values by direct method or iterative method.

Above the BCs are derived from physical observation. Mathematically, the BC can be derived as follows:

We know $v = -\partial\psi/\partial x = 1$ between point $\psi_{1,1}$ and $\psi_{2,1}$. To follow easier, let us call $\psi_{1,1}$ as point 1 and $\psi_{2,1}$ as point 2.

Rewriting $\partial\psi/\partial x = -1$ or $\partial\psi = -\partial x$ and integrating both sides, one get $\boldsymbol{\psi = -x}$, with limits

$$\int_1^2 \partial\psi = -\int_1^2 \partial x$$

$$\psi_2 - \psi_1 = -(x_2 - x_1) = x_1 - x_2,$$

Hence, the ψ values vary linearly between the points 1 and 2 for constant velocity. If the velocity is not constant, the integration will give different variation.

Knowing $x_1 = 0$ and $x_2 = 1$ and keeping $\psi_1 = 1$, one gets $\psi_2 = 0$. For any other interior points, one can linearly interpolate.

3.3.3 Direct Method of Solution

Knowing the boundary values all around for ψ, we can calculate interior values by forming simultaneous equations. For ease of writing the equations in the matrix form, let us represent the interior point or unknowns as 1 to 4, as shown in Figure 3.3.

Applying the equation (3.1) into four points, one gets:

Point 1 : $\quad 4\psi_1 - \psi_2 - \psi_4 - 1 - 0 = 0 \quad 4\psi_1 - \psi_2 - \psi_4 = 1$

Point 2 : $\quad -\psi_1 + 4\psi_2 - \psi_3 - 1 - 0 = 0 \quad -\psi_1 + 4\psi_2 - \psi_3 = 1$

Point 3 : $\quad -\psi_2 + 4\psi_3 - \psi_4 - 1 - 1 = 0 \quad -\psi_2 + 4\psi_3 - \psi_4 = 2$

Point 4 : $\quad -\psi_1 - \psi_3 + 4\psi_4 - 1 - 1 = 0 \quad -\psi_1 - \psi_3 + 4\psi_4 = 2$

(a)

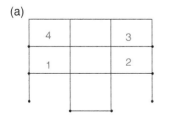

(b)

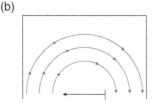

Figure 3.3 (a) Grid with numbering to form matrix equations (b) stream lines to show the symmetry about the vertical axis.

The preceding equation can be written in matrix form Ax = b as shown later:

$$
\begin{bmatrix} 4 & -1 & 0 & -1 \\ -1 & 4 & -1 & 0 \\ 0 & -1 & 4 & -1 \\ -1 & 0 & -1 & 4 \end{bmatrix} \begin{bmatrix} \psi_1 \\ \psi_2 \\ \psi_3 \\ \psi_4 \end{bmatrix} = \begin{bmatrix} 1 \\ 1 \\ 2 \\ 2 \end{bmatrix}
$$

Above Ax = b equation can be solved by Gaussian elimination or Cholesky decomposition. In this problem, symmetry is noticed about the y-axis and hence substitute $\psi_1 = \psi_2$ and $\psi_3 = \psi_4$. Then the equation becomes:

$$
\begin{bmatrix} 3 & -1 \\ -1 & 3 \end{bmatrix} \begin{bmatrix} \psi_1 \\ \psi_3 \end{bmatrix} = \begin{bmatrix} 1 \\ 2 \end{bmatrix}
$$

Multiplying second equation by 3 and summing the two equations:

$$8\psi_3 = 7,$$

Hence, $\psi_3 = 7/8 = 0.875$.
Substituting in the first equation ψ_3 one gets:

$$3\psi_1 - \psi_3 = 1, 3\psi_1 = 1.875, \psi_1 = 0.625.$$

From symmetry, we get: $\psi_1 = \psi_2 = 0.625$ and $\psi_3 = \psi_4 = 0.875$.
For problems to solve by direct solution takes lots of computer time when the number of unknowns is increased.

3.3.4 Memory Requirements for a 100 × 100 Mesh

For example, if one has 100x100 mesh, then we can say the number of equations is 10 000 including the boundaries. Then the size of the A matrix is $10^4 \times 10^4$, and the number of elements of the A matrix is 10^8. The memory required for the A matrix assuming 8 bytes for each storage location is 8×10^8 bytes or 800 Mbytes. If symmetry of the matrix is utilized, then the memory for A matrix may be 400 Mbytes. If banded solver is used with semi band-width of 100, the memory will be $8 \times 100 \times 10^4 = 8 \times 10^6$ bytes or 8 Mbytes. For detail of the storage, one can refer to Figure 3.17. By using iterative technique, we will show that the memory requirement for A matrix is almost none. If variable grid spacing is used, then it may be 5 coefficients for each equation and hence 5×10^4 size of the array is needed. The memory requirements for the A matrix are 4×10^5 bytes or 0.4 Mbytes. For 3D problems, the memory saving is many 100's of times.

3.3.5 Iterative Method by Gauss–Siedel (GS) or Successive Over Relaxation (SOR)

The following algorithm illustrated the Gauss–Siedel (GS) and successive over relaxation (SOR) method.

DO j = 2, 3
DO i = 2, 3
$$RES = 4\psi_{i,j} - \left(\psi_{i+1,j} + \psi_{i-1,j} + \psi_{i,j+1} + \psi_{i,j-1} \right)$$
$$\psi_{i,j} = \psi_{i,j} - RES^*RF/4$$
END DO − i
END DO − j

Repeat the preceding calculations, until the residue (RES) becomes less than tolerance in all the equations.

If relaxation parameter RF = 1, then the iterative method is called Gauss–Siedel (GS) method and if RF > 1, it is called successive over relaxation (SOR) method. Both the methods of solving Ax = b are illustrated using Excel sheet as if it is done in the computer models. Here initially all the interior values are started with zero.

Iteration-GS	s11	s21	s31	s41	s12	s22	s32	s42	s13	s23	s33	s43	s14	s24	s34	s44
Starting	1	0	0	1	1	0	0	1	1	0	0	1	1	1	1	1
1	1	0	0	1	1	0.25	0.3125	1	1	0.578125	0.722656	1	1	1	1	1
2	1	0	0	1	1	0.472656	0.548828	1	1	0.817871	0.841675	1	1	1	1	1
3	1	0	0	1	1	0.591675	0.608337	1	1	0.862503	0.86771	1	1	1	1	1
4	1	0	0	1	1	0.61771	0.621355	1	1	0.872266	0.873405	1	1	1	1	1
5	1	0	0	1	1	0.623405	0.624203	1	1	0.874402	0.874651	1	1	1	1	1
6	1	0	0	1	1	0.624651	0.624826	1	1	0.874869	0.874924	1	1	1	1	1
7	1	0	0	1	1	0.624924	0.624962	1	1	0.874971	0.874983	1	1	1	1	1
8	1	0	0	1	1	0.624983	0.624992	1	1	0.874994	0.874996	1	1	1	1	1
SOR-RF = 1.5	s11	s21	s31	s41	s12	s22	s32	s42	s13	s23	s33	s43	s14	s24	s34	s44
Starting	1	0	0	1	1	0	0	1	1	0	0	1	1	1	1	1
1	1	0	0	1	1	0.325	0.430625	1	1	0.789953	1.046688	1	1	1	1	1
2	1	0	0	1	1	0.624188	0.738847	1	1	0.993313	0.898946	1	1	1	1	1
3	1	0	0	1	1	0.700696	0.623229	1	1	0.846713	0.858048	1	1	1	1	1
4	1	0	0	1	1	0.592523	0.609466	1	1	0.872928	0.874364	1	1	1	1	1
5	1	0	0	1	1	0.629022	0.63076	1	1	0.877287	0.877806	1	1	1	1	1
6	1	0	0	1	1	0.626409	0.624642	1	1	0.87511	0.874077	1	1	1	1	1

To illustrate the GS and SOR by hand calculations, we can take the preceding two equations as follows:

$$R1 = 3\psi_1 - \psi_3 - 1$$
$$R2 = -\psi_1 + 3\psi_3 - 2$$

Here R1 and R2 are residues for equations 1 and 2. The procedure for hand calculations is

1) Assume initial values for ψ_1 and ψ_3 to be zero if it is not known or estimated.
2) Calculate the residue for each equation knowing the current values. Update the values by $\psi_i = \psi_i - R_i \times RF/ap$. Where ap is the diagonal value for that equation and in this case, they are 3.
3) Repeat the calculation until the **residue reduced to required accuracy**.

For GS alone, we can solve directly from the following equations if we are not interested in the residue:

$$\psi_1 = (1 + \psi_3)/3 \quad \text{and} \quad \psi_3 = (2 + \psi_1)/3$$

In the following tables, both residue and solution are provided for each iteration.

GS calculations-RF = 1

Iteration#	1	2-Res	2-ψ	3-Res	3-ψ	4-Res	4-ψ	Exact-ψ
ψ_1	0	−1	0.333	−0.778	0.593	−0.085	0.621	0.625
ψ_3	0	−2.333	0.778	−0.259	0.864	−0.029	0.874	0.875

More GS iterations are done it will converge to the exact solution. Also, note that the residues are decreasing as the iteration decreases.

SOR calculations-RF = 1.5

Iteration#	1	2-Res	2-ψ	3-Res	3-ψ	4-Res	4-ψ	Exact-ψ
ψ_1	0	−1	0.5	−0.75	0.875	0.812	0.469	0.625
ψ_3	0	−2.5	1.25	0.875	0.813	−0.0313	0.828	0.875

In this case, the solution is not converging consistently. May be RF = 1.5 is not the optimum. The same RF = 1.5 for the four preceding equation gave faster convergence. From numerical experiments, it is found that if one uses RF = 1.05, one gets $\psi_1 = 0.626$ and $\psi_3 = 0.875$.

The convergence criteria for GS or SOR iterative method are not discussed here. But for detail one can refer to Patankar (1980).

3.3.6 Details of Program Pcham.f.

Program **pcham.f** can calculate the stream function ψ or S as a variable in the program for all the interior points using SOR. The program has the flexibility of having inlet and outlet as specified by the user. As an example, let us consider the same 3 units × 3 units square region above with more points. Keeping grid spacing h = 0.25, we will have number of points in the x and y directions (IM, JM) as 13 and 13. All around, the S value is 1 except on the bottom region.

On the bottom region (along J = 1), I = 1 to 5 S varies linearly in the **descending order**: 1.0, 0.75, 0.5, 0.25, 0. Similarly for I = 9 to 13, S varies in the **ascending order**: 0, 0.25, 0.5, 0.75, 1.0. The grid and the values on the boundary can be seen in Figure 3.4a. Knowing stream function, the velocities u and v are calculated for the interior points as follows:

$$u = \partial\psi/\partial y \quad \text{and} \quad v = -\partial\psi/\partial x$$

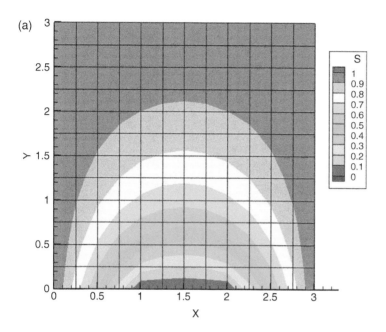

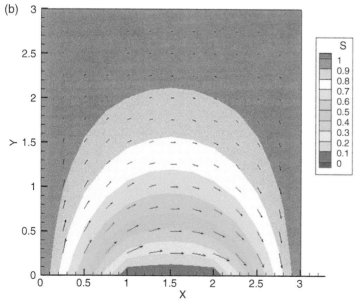

Figure 3.4 (a) Grid, boundary condition and interior stream function values for the flow in a chamber problem (b) velocity vector plot.

The input details in **pcham-i.txt** are:

```
Line1:        READ(2,*)IM,JM,H,RF
IM    Number of points in the X-direction
JM    Number of points in the y-direction
H     length of grid spacing
RF    Relaxation factor for GS or SOR


c.....READ BOUNDARY VALUES
Line2:    READ(2,*)(S(I,1),I=1,IM)
Give IM points of stream function values for bottom boundary

Line3:    READ(2,*)(S(I,JM),I=1,IM)
Give IM points of stream function values for top boundary

Line4:    READ(2,*)(S(1,J),J=2,JM-1)
Give interior points (JM-2 points) of stream function values for
left boundary

Line5:    READ(2,*)(S(IM,J),J=2,JM-1)
Give interior points (JM-2 points) of stream function values for
right boundary

Input data-pcham-i.txt:
13,13, 0.25, 1.5
1.0,0.75,0.5,0.25,0, 0,0, 0, 0,0.25,0.5,0.75,1.0 (13 values-bottom
boundary)
1.,1.,1.,1.,1., 1.,1.,1.,1.,1., 1.,1.,1. (13 values-top boundary)
1.,1.,1.,1.,1.,  1.,1.,1.,1.,1.,   1.   (11 values-left boundary
interior points)
1.,1.,1.,1.,1., 1.,1.,1.,1.,1., 1. (11 values-right boundary
interior points)
```

Output details: The program writes the final values of S, u, and v in **pcham-p.plt** for tec-plot visualization. The number of iteration and the error for each iteration S are written on the screen.

Tolerance for Stopping the Iterative Method: The iterative method is repeated until required accuracy of the solution is achieved. There are several methods for measuring the error norm. The ideal one is to run the iteration until all the equation residues reduce to zero. For most of the engineering problems, that level of accuracy is not needed. The following criteria can be used:

1) Absolute sum of the residue for all the equations $< \text{TOL} = \text{IM}^*\text{JM}^*10^{-5}$. Here the average error in each equation is kept less than 10^{-5}.
2) Absolute maximum error of all the equations $< \text{TOL} = 10^{-5}$. Here the maximum error is less than 10^{-5}. This may be a stringent criterion than error norm in point 1.

In the preceding statement, error in each equation is less than 10^{-5}. What is the proper number to be used that depends upon several factors such as the equations are nondimensional or dimensional. Depending upon the problem, one has to choose it from experience.

3.3.7 Optimum Relaxation Parameter RF for SOR

For the above data with RF = 1.5, the program took 45 iterations to converge. By changing the RF values, one can see how the RF influences the total number of iterations, as shown in Figure 3.5. In this case, the optimum relaxation factor came to be 1.6, with total number of iterations for convergence as 26. The residual error for each iteration for RF = 1.6 is shown in Figure 3.6. Hence, in this case, the SOR with RF = 1.6 is 5.3 times faster than GS method.

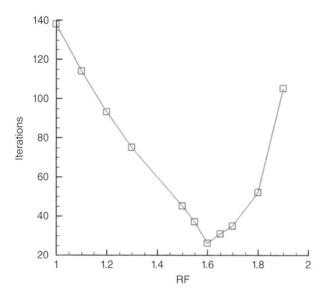

Figure 3.5   Optimum relaxation factor RF for 13 × 13 grid.

```
Command Prompt                                                            —   □   ×

C:\selvam\sel-20\department\cwe-book\ch3>pcham
         1     90.250942128065986
         2     39.485488726984613
         3     27.510148247255003
         4     21.414456706770267
         5     17.202077003455216
         6     13.840669219557983
         7     11.184410277579413
         8      9.0522514063985078
         9      7.3271156625304572
        10      5.9170092762663407
        11      4.8024095455327762
        12      3.6202306639911668
        13      2.5667469850990932
        14      1.7621865832398664
        15      1.1709678825839227
        16      0.75845167485351483
        17      0.47187069346755639
        18      0.27693852101240751
        19      0.15818642179579209
        20      9.0567567936762544E-002
        21      5.3237574820425439E-002
        22      4.3615724344092603E-002
        23      5.2269268902010932E-002
        24      8.2900996685944389E-003
        25      4.8174475915097222E-003
        26      1.4398301939472047E-003
C:\selvam\sel-20\department\cwe-book\ch3>
```

Figure 3.6   Screen run shows the number of iterations for RF = 1.6.

If the solver is used many times for a particular grid as in the case of unsteady problem, then finding the optimum RF may help to solve a problem very efficiently with less computer time.

Program pcham.f listing:

```
c       PROG. PCHAM.F, MAY 17, 2020
c       FLOW IN A CHMABER BY POTENTIAL FLOW
c       CVEG 5633-CWE class, R.P. Selvam, email: rps@uark.edu
        PARAMETER (NX=51,NY=51)
        IMPLICIT REAL*8 (A-H,O-Z)
        DIMENSION S(0:NX+1,0:NY+1), U(NX,NY),V(NX,NY)
c.....OPEN FILE FOR TECPLOT
        OPEN(2,FILE='pcham-i.txt')
        OPEN(7,FILE='pcham-p.plt')
        READ(2,*)IM,JM,H,RF
        TOL=IM*JM*1.E-5
        NITER=900
c.....RELAXATION FACTOR GS:RF=1; SOR OR OVER-RELAXATION RF>1.0
c       & UNDER-RELAXATION RF<1.0
c        RF=1.0
c.....IMPLEMENT BC'S & INTIAL CONDITIONS
        DO J=1,JM
        DO I=1,IM
        S(I,J)=0.0
        U(I,J)=0.0
        V(I,J)=0.0
        END DO
        END DO
c.....READ BOUNDARY VALUES
        READ(2,*)(S(I,1),I=1,IM)
        READ(2,*)(S(I,JM),I=1,IM)
        READ(2,*)(S(1,J),J=2,JM-1)
        READ(2,*)(S(IM,J),J=2,JM-1)
c.....SOLVE BY GAUSS SIEDEL OR SOR
        AP=4.0
        ITER=0
1000    ITER=ITER+1
        RNORM=0.0
        DO J=2,JM-1
        DO I=2,IM-1
        RES=AP*S(I,J)-(S(I+1,J)+S(I-1,J)+S(I,J+1)+S(I,J-1))
        RNORM=RNORM+ABS(RES)
        S(I,J)=S(I,J)-RF*RES/AP
        END DO
        END DO
```

```
       PRINT *,ITER,RNORM
       IF(ITER.LT.NITER.AND.RNORM.GT.TOL)GO TO 1000
c.....Calculate U AND V
       DO J=2,JM-1
       DO I=2,IM-1
       U(I,J)=(S(I,J+1)-S(I,J-1))/(2.0*H)
       V(I,J)=-(S(I+1,J)-S(I-1,J))/(2.0*H)
       END DO
       END DO
c.....WRITE FOR TECPLOT
       write(7,*)'VARIABLES = "X","Y","S","U","V"'
       write(7,*)'ZONE I=',IM, ',J=',JM,',F=POINT'
       do j=1,jm
       do i=1,im
       X=(I-1)*H
       Y=(J-1)*H
       write(7,*)X,Y,S(I,J),U(I,J),V(I,J)
       end do
       end do
       STOP
       END
```

3.3.8 Inviscid Flow Over a Square Cylinder or Building

By modifying the preceding program, one can compute the flow over a square cylinder of dimensions 1unit × 1unit. This problem looks similar to flow over a building, and the details of the computational region and BCs are shown in Figure 3.7. Because of symmetry about x-axis, only upper portion is considered. If the vertical symmetry was considered, the computation would have reduced. For visualization purposes, the other symmetry is not considered. To consider the effect of building or cylinder in the flow, the points inside the cylinder are not considered for computation. To introduce this effect, the number of points to be omitted for computation along the vertical axis from J = 1 needs to be entered along the x-axis, and this array is called KHI(I = 1 to IM). Here it is assumed that the stream function is zero at the bottom and goes over the building. The flow region is 8 × 4 units.

Figure 3.7 Flow over a square cylinder or building.

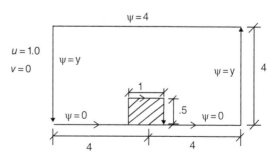

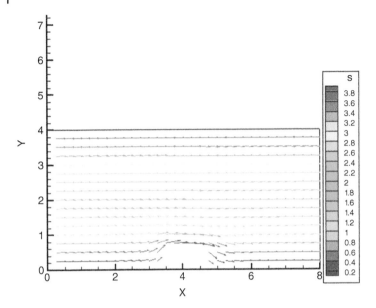

Figure 3.8   Velocity vector around the cylinder.

The flow goes from left to right with u = 1.0 and v = 0. Using these values, one can calculate the boundary condition at the left and right boundary to be ψ = y and the top boundary ψ = y = 4. To illustrate the application, h = 0.25 unit is considered. The cylinder will have 4 × 2 grid spacing. The domain will have 33 × 17 points. From I = 15 to I = 19, one gives KHI = 3 and the rest as 1. The velocity vector plot is reported in Figure 3.8. The maximum u and v velocities are 1.431 and 0.673 units, respectively. One can see that there is no flow separation for the inviscid flow. For viscous flow, there will be flow separation and that will be discussed in Chapter 5.

The same computer program can be used for flow over circular cylinder if a greater number of points are considered to approximate the circular region. Then one can compare the results with closed-form solution reported in any fluid mechanics book.

3.3.9   Iterative Solvers Used in Practical Applications

We only discussed GS and SOR iterative solvers. For practical applications, preconditioned conjugate gradient (PCG), multigrid (MG) and MGCG solvers are used. They give solutions for complex problems in reasonable number of iterations (say less than 100 iterations). The SOR type solver can takes many thousands of iterations. The algorithm for PCG solver with ILU preconditioner in a structured grid is reported in Selvam (1996). For the MGCG solver, MG is used as a preconditioner for the PCG. The MGCG is used by Sarkar and Selvam (2009) for spray cooling application. For more detail, one can refer to Ferziger and Peric (2002) and Selvam (2020).

In the problems discussed so far, the specified BC is ψ or Dirichlet. For this kind of problems, the iterative method converges faster than when ∂ψ/∂n is specified as BC. As we discussed this kind is called Neumann BC. This is the type of BC one has when one solves for pressure in the NS equations.

Summary for Steady-State Solutions: One solves the simultaneous equations one time for steady-state solution. So, any physical problem that can be modeled as a steady state, then that is the fastest solution. In the wind engineering area, mostly we faced with unsteady problems. Same way if a problem can be modeled as 1D or 2D, they take less computer time than 3D. On the other hand, most of the wind engineering applications are 3D problems. To understand the numerical solution issues, we will consider a 1D problem in the next section.

3.4 Unsteady Problem-Explicit and Implicit Solution for the Wave Equation

For physical understanding of convection or advection phenomena, let us consider the approaching of a shock wave or tsunami wave shown in https://www.sms-tsunami-warning.com/pages/tsunami-drawback#.Ybjmz73MI2w. When the wave interacts with a structure, the pressure on the structure changes in time as shown in the Figure 3.9. Similarly, due to turbulence in wind as we discussed in Chapter 2, the velocity changes in time. The turbulence in wind can be represented by a sum of sine and cosine series using Fourier analysis as illustrated in Chapter 2. Hence, transport of a single sine wave using advection equation will explain the challenges in modeling advection equation.

To illustrate this unsteady problem using convection or advection equation, let us consider the following one-dimensional (1D) wave equation:

$$dA/dt + UdA/dx = 0.0 \quad \text{in } 0 < x \leq 2L \tag{3.2}$$

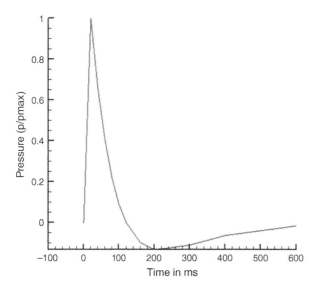

Figure 3.9 Shock wave.

With an initial condition at t = 0 as, A(x, 0) = sin(2πx/L) and the BCs are at x = 0, A(0, t) = sin(−2πt/L) and at x = 2L, **convective boundary condition** is used, i.e., in this case Equation (3.2).

The exact solution to the above problem is:

$$A(x, t) = \sin\left[2\pi(x - t)/L\right] \tag{3.3}$$

Here A is the amplitude, L is the wavelength, and the sine wave travels at a speed of 1 unit (U = 1). This problem is a linear problem, and it is a good benchmark problem to understand the challenges in approximating the convection part of the NS equations. When we plot in time, the amplitude changes at different time for a particular point in x. Hence, this is a space and time problem. As an example, we plot the Equation (3.3) for a domain length of 2 units with wavelength L = 1 unit at time t = 0 and at t = 0.25, as shown in Figure 3.10. From the figure, it is clearly seen as time changes the amplitude changes at different points. This is what happens in transporting turbulence in the wind. To predict these amplitudes, we will use FDM approximations in time and space.

The first term dA/dt is approximated in time by FDM. Generally, current (n + 1) and old time (n) steps are considered. Only in the Adam–Bashforth method discussed later n + 1, n, and n − 1 time steps are considered. The term dA/dx is differentiated in space, but this is approximated in time and space. That is dA/dx in space is approximated by considering time n or n + 1 or both. Depending upon these approximations in time and space, different methods evolve as discussed later.

Selvam (1998) used a 2D benchmark problem to evaluate the performance of different methods. In the NS equation, the convection term is nonlinear, and further numerical complications are encountered. From Fourier analysis, we saw that minimum 3 points are needed to represent a sine or cosine wave. But using FDM, the error will be high, as shown in Figure 3.11a. The preceding 1D problem is used by Mansouri et al. (2020) to assess how

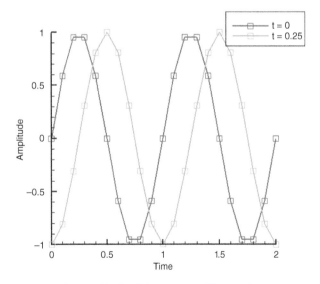

Figure 3.10 Amplitude of the wave at different times.

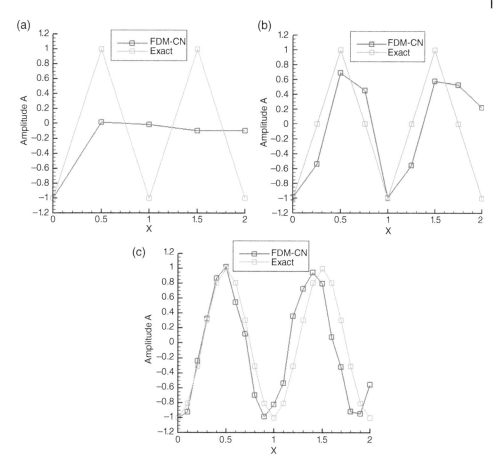

Figure 3.11 Comparison of an exact sine wave with the FDM method after two cycles of transport. (a) Wavelength L=2h (3 points/wave), (b) wavelength L=4h (5 points/wave), and (c) wavelength L=10h (11 points/wave).

many points are needed to transport a sine wave with reasonable accuracy. They suggested minimum 5 points, but the error is substantial, as shown in Figure 3.11b. Normally, 10 grid spacings (10h) are needed, and we will evaluate this in this section. When we transport, we like to make sure that any particle does not travel more than a grid spacing. That is the time step **dt should be less than h/U (dt < h/U)**, where U is the wave velocity, and it is equal to one here. This criterion is called Courant–Friedrichs–Lewy (CFL) condition. The **CFL condition is specified to be (Udt/h < 1)**. We can evaluate the performance of different methods for different CFL conditions.

This problem will demonstrate various issues as follows:

1) Importance of using explicit and implicit solution methods. Here we only use, Euler backward implicit and Euler forward explicit, Crank–Nicolson (CN) implicit method and Adam–Bashforth (AB) explicit method. The last two are second-order accuracy.

2) Several methods are available in the literature, but only these four are considered here. For other methods and for further understanding one can refer to Tannehill et al. (1997). Even today, there is no one method that can be used for every problem, and it is an open ground for further research.

3) The implicit method-like CN guaranties error reduction to required accuracy at each time step. Only part is, it needs a simultaneous solver of the kind Ax = b and hence it is computationally demanding and computer storage is high. For 2D and 3D general problems, still not clear what kind of solution techniques to use.

4) The AB method is computationally efficient because there is no need to solve a simultaneous equation as in CN. How well it satisfies the equation is not clear. May be with the current benchmark problem we evaluate. Spectral method group extensively uses this method because of its ease in the implementation. The other drawbacks are CFD < 1 (condition to satisfy for convergence) and constant time step requirement. Orlandi (2000) also used this method for his FDM applications.

5) The partial differential equation is in time and space. For space, central difference or upwinding can be used. The central difference method is second-order accurate and for turbulent transport problem CD method is preferred compared to upwinding methods. The upwinding method is diffusive, and CD method is dispersive. For LES simulation CD is preferred. For certain class of problems like k–ε Reynolds averaged turbulence model, QUICK type second-order upwinding is used.

6) This wave equation is a linear problem, but the convection part of NS equation is nonlinear. There is no accurate one method is available to use. Depending upon the problem, one has to choose the method from experience. There are several methods in use and only few is used for illustration.

3.4.1 Discretization of the Wave Equation by Different FDM Schemes

Explicit Procedure:

Euler Forward Upwinding Method (EFUM): In this procedure, since traveling wave velocity is positive, the space is approximated by backward difference. In the upwinding procedure if the velocity is negative, then forward procedure has to be used. For more explanation one has to look at Patankar (1980) book. The superscript "n" is for time and subscript "i" is for space position. Then Equation (3.2) is approximated by finite difference as follows:

$$\left(A_i^{n+1} - A_i^n\right)/dt + \left(A_i^n - A_{i-1}^n\right)/h = 0.0$$

Rewriting $A_i^{n+1} = A_i^n - \left(A_i^n - A_{i-1}^n\right)dt/h$ for i = 2 to IM

The preceding method is stable if CFL is kept less than 1. For the convection problem using Euler forward procedure for stability reason, only upwinding (UW) should be used. For details of general implementation of UW, one can refer to Patankar (1980) or Selvam (2020)

Adam–Bashforth Method (ABM): In the Adam–Bashforth method, the space variable is approximated using n and n−1 time step as follows:

$$\left(A_i^{n+1} - A_i^n\right)/dt + \left(3f^n - f^{n-1}\right)/2 = 0.0 \text{ for } i = 2 \text{ to IM} - 1 \text{ and } f = dA/dx$$
$$= (A_{i+1} - A_{i-1})/2h$$

Rewriting to solve for A at time $n+1$:

$$A_i^{n+1} = A_i^n - \left(3f^n - f^{n-1}\right) dt/2 \text{ for } i = 2 \text{ to IM} - 1$$

For $i = $ IM, $A_i^{n+1} = A_i^n - \left(3f^n - f^{n-1}\right) dt/2$ where $f = dA/dx = (A_i - A_{i-1})/h$

For the last point, we use UW.

In the preceding derivation, we used CD for space approximation for $I = 2$ to IM-1. Here UW can also be used but the diffusive error may be high.

Balanced Tensor Diffusivity (BTD) Scheme or Leith Method: The Leith method is a second-order explicit method. For 2D and 3D applications, one can refer to Selvam (2020) notes. Here the wave equation is modified to:

$$dA/dt + UdA/dx - dtU^2d^2A/dx^2/2 = 0.0$$

Using $U = 1$, the FDM version is

$$\left(A_i^{n+1} - A_i^n\right)/dt + \left(A_{i+1}^n - A_{i-n}^n\right)/2h - dt\left(A_{i+1}^n - 2A_i^n + A_{i-1}^n\right)/2h^2$$
$$= 0.0 \quad \text{for } I = 2 \text{ to IM} - 1$$

Rewriting : $A_i^{n+1} = A_i^n - dt\left(A_{i+1}^n - A_{i-n}^n\right)/2h + dt^2\left(A_{i+1}^n - 2A_i^n + A_{i-1}^n\right)/2h^2$
for $I = 2$ to IM $- 1$

For IM point, we use UW as in ABM. This method will be better than EFUM.

In all the preceding methods, the values for the current time are calculated explicitly from the previous time values. In other words, no simultaneous equations are needed. Hence, computationally these methods are faster. Whereas in the implicit methods discussed later, one has to solve simultaneous equations at each time.

Implicit Procedure:

In the implicit procedure, the space can be approximated by CD or UW method, and the method is stable. Here we only consider CD method. Even CFL > 1 can be used for practical purposes, only CFL < 1 will be used for illustration.

Backward Euler Method (BEM): In this method, the space is approximated in the $n + 1$ time step as shown later.

$$\left(A_i^{n+1} - A_i^n\right)/dt + \left(A_{i+1}^{n+1} - A_{i-1}^{n+1}\right)/2h = 0.0 \text{ for } i = 2 \text{ to IM} - 1$$

Assembling proper coefficients, we can write in the form:

$$RES = AP^*A_i^{n+1} + AE^*A_{i+1}^{n+1} + AW^*A_{i-1}^{n+1} + SC$$

where $AP = 1/dt$, $AE = 1/2h$, $AW = -1/2h$ and $SC = -A_i^n/dt$.

For the boundary point IM, UW can be used for space. This reduces to:

$$\left(A_i^{n+1} - A_i^n\right)/dt + \left(A_i^{n+1} - A_{i-1}^{n+1}\right)/h = 0.0$$

The preceding simultaneous equation can be solved by Gauss–Siedel (GS) or line solver-like TDMA (Tri Diagonal Matrix Algorithm). The TDMA implantation is provided in Patankar (1980) or Selvam (2020). In our applications here, we will use GS solver.

Crank–Nicolson Method (CNM): In the CN, the space is averaged between n and $n + 1$ time steps.

$$\left(A_i^{n+1} - A_i^n\right)/dt + \left[\left(A_{i+1}^{n+1} - A_{i-1}^{n+1}\right)/2h + \left(A_{i+1}^n - A_{i-1}^n\right)/2h\right]/2 = 0.0 \text{ for } i = 2, IM-1$$

The coefficients will become: $AP = 1/dt$, $AE = 1/4h$ and $AW = -1/4h$ and $SC = -A_i^n/dt + \left(A_{i+1}^n - A_{i-1}^n\right)/4h$

For the boundary point IM, UW can be used for space. This reduces to

$$\left(A_i^{n+1} - A_i^n\right)/dt + \left[\left(A_i^{n+1} - A_{i-1}^{n+1}\right)/h + \left(A_i^n - A_{i-1}^n\right)/h\right]/2. = 0.0$$

To illustrate the implementation of the preceding algorithms into program, program **w1dab.f** and **w1dcn.f** based on ABM and CNM, respectively, are listed later. The **w1dab.f** is based on the explicit method and **w1dcn.f** is based on the implicit method. The other methods (UW and BTD) can be easily implemented by modifying the two programs and hence they are not reported (**w1duw.f** and **w1dbtd.f**).

Program w1dab.f:

```
c      PROGRAM W1DAB.F, 7/19/2020
c      WAVE 1D USING ADAM-BASHFORTH
c      PROGRAM W1DCN.F, 7/19/2020
c      PROGRAM WAVE1D.F, 6/17/2020
c      TRANSPORT OF 1d WAVE FOR L=2H & 4H USING cn-cd.
c      THE DOMAIN LENGTH 2L & CFL=0.25
       PARAMETER (NX=101)
       IMPLICIT REAL*8 (A-H,O-Z)
       DIMENSION S(NX,3),SF(NX)
       OPEN(2,FILE='w1dab.plt')
       PHI=4.*ATAN(1.0)
c       IM=5
c       XL=2.0
c       CFL=0.1
c       TTIME=2.25
       PRINT *,'GIVE IM,XL,CFL,TTIME'
       READ(*,*)IM,XL,CFL,TTIME
c
       TOL=IM*1.E-5
       NITER=200
       H=XL/(IM-1)
       H2=2.*H
       DT=CFL*H
c.....INITIALIZE
       DO I=1,IM
       X=(I-1)*H
```

```
      S(I,1)=SIN(2.*PHI*X)
      S(I,2)=S(I,1)
      S(I,3)=SIN(2.*PHI*(X+DT))
      END DO
c.....START TIME SOLUTION
      TIME=0.0
      ITERT=0
1000  ITERT=ITERT+1
      TIME=TIME+DT
c.....SOLVE FOR NEXT TIME
      S(1,1)=SIN(-2.*PHI*TIME)
c
      DO I=2,IM-1
      FN=(S(I+1,2)-S(I-1,2))/H2
      FNM1=(S(I+1,3)-S(I-1,3))/H2
      S(I,1)=S(I,2)-(3.*FN-FNM1)*DT/2.
      END DO
c.....IMPOSE BC
      S(IM,1)=S(IM,2)-DT*(S(IM,2)-S(IM-1,2))/H
c.....REPLACE OLD TO NEW
      DO I=1,IM
      S(I,3)=S(I,2)
      S(I,2)=S(I,1)
      END DO
      IF(TIME.LT.TTIME)GO TO 1000
c.....WRITE OUTPUT IN A FILE FOR PLOTTING
      DO I=1,IM
      X=(I-1)*H
      AVALUE=SIN(2.*PHI*X-2.*PHI*TTIME)
      WRITE(2,*)X,S(I,1),AVALUE
      END DO
      STOP
      END

Program w1dcn.f:
c     PROGRAM W1DCN.F, 7/19/2020
c     PROGRAM WAVE1D.F, 6/17/2020
c     TRANSPORT OF 1d WAVE FOR L=2H & 4H USING cn-cd.
c     THE DOMAIN LENGTH 2L & CFL=0.25
      PARAMETER (NX=101)
      IMPLICIT REAL*8 (A-H,O-Z)
      DIMENSION S(NX,2),SF(NX)
      OPEN(2,FILE='w1dcn.plt')
      PHI=4.*ATAN(1.0)
```

```
c       IM=5
c       XL=2.0
c       CFL=0.1
c        TTIME=2.25
        PRINT *,'GIVE IM,XL,CFL,TTIME'
        READ(*,*)IM,XL,CFL,TTIME
c
        TOL=IM*1.E-5
        NITER=200
        H=XL/(IM-1)
        H4=4.*H
        DT=CFL*H
        AP=1./DT
        AE=1./H4
        AW=-AE
c.....INITIALIZE
        DO I=1,IM
        X=(I-1)*H
        S(I,1)=SIN(2.*PHI*X)
        S(I,2)=S(I,1)
        END DO
c.....START TIME SOLUTION
        TIME=0.0
        ITERT=0
1000    ITERT=ITERT+1
        TIME=TIME+DT
c.....SOLVE FOR NEXT TIME
        S(1,1)=SIN(-2.*PHI*TIME)
        ITER=0
1100    ITER=ITER+1
        RNORM=0.0
        DO I=2,IM-1
        SC=(S(I+1,2)-S(I-1,2))/H4-S(I,2)/DT
        RES=AP*S(I,1)+AE*S(I+1,1)+AW*S(I-1,1)+SC
        RNORM=RNORM+ABS(RES)
        S(I,1)=S(I,1)-RES/AP
        END DO
c.....IMPOSE BC
        S(IM,1)=S(IM,2)-DT*(S(IM,1)-S(IM-1,1))/H
        IF(RNORM.GT.TOL.AND.ITER.LT.NITER)GO TO 1100
        print *,time,iter,rnorm
c.....REPLACE OLD TO NEW
        DO I=1,IM
        S(I,2)=S(I,1)
        END DO
        IF(TIME.LT.TTIME)GO TO 1000
```

```
c.....WRITE OUTPUT IN A FILE FOR PLOTTING
      DO I=1,IM
      X=(I-1)*H
      AVALUE=SIN(2.*PHI*X-2.*PHI*TTIME)
      WRITE(2,*)X,S(I,1),AVALUE
      END DO
      STOP
      END
```

3.4.2 Input Preparation

We will consider L = 1 and hence the total length of the domain is 2 for the problem stated in Equation (3.2). We will use 10 spacing per wave and this amounts to IM = 21 points and h = 2./(IM-1) = 0.1. We will keep dt less than 0.1 units to have CFL (Udt/h = dt/h < 1) less than 1. We will run for a total time (TTIME) more than 2-time units.

The CFL relates to time step dt. If CFL is far less than one, then dt will be smaller. Smaller dt means more number of time steps to cover the specified total time to run. As an example, CFL = 0.5 means dt = 0.5h = 0.05. Hence to run 2-time units, one needs to run 40-time steps. If CFL = 0.25, dt = 0.025 and hence the number of time step increased to 80. With respect to accuracy, lesser time step will give less error. On the other hand, lesser time step takes a greater number of time step and hence more computer time.

Same way, the IM refers to number of grid points for the given domain. If IM increases, the spatial accuracy is increased and also the size of the A matrix. So, in practical problems, one has to make judicial choice on number of points in space and the time step.

Input: We will give the following input in the DOS screen for the programs:

```
IM,XL,CFL,TTIME
IM       # of points in the x-direction
XL       Length of the computational domain
CFL      CFL number
TTIME    Total time to run
```

Sample data: 21, 2, 0.5, 2.

Internal Variable Details: Max. storage for the variable: 101, Max. # of GS iteration allowed: 500, Tolerance: 10^{-5}IM.

Screen Output: In the screen at each time step, the following values are written:

Time, # of GS iteration, absolute sum of the residue of all the equations.

```
Output file:  w1dcn.plt or w1dab.plt,
Contains: WRITE(2,*)X,S(I,1),AVALUE for I=1 to IM
x          distance
S(I,1)     Amplitude from computation
AVALUE     Amplitude from exact solution
```

The final wave profiles are plotted for CN, AB, UW, and BTD methods, as shown in Figure 3.12. From the figure, we can see that AB method has better performance than

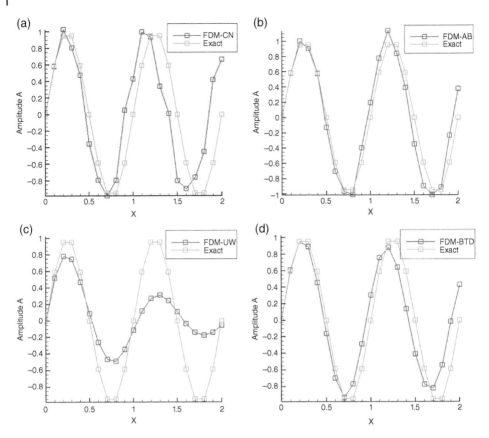

Figure 3.12 Wave profile after 2-time units. (a) Using CN method, (b) using AB method, (c) using UW method, (d) using BTD or Leith scheme.

CN method. Further analysis can be done by varying the CFL and TTIME. The dispersive error using CN method is reduced in AB method. The other advantage is, it takes less CPU because of explicit method. Only drawback is, it needs one more memory space for the variable.

The UW scheme is very diffusive, but BTD scheme is better than upwind. Still, BTD is diffusive some level. For 2D or 3D application problem, we can try both CD and BTD scheme in the later sections.

By increasing the number of grid points or spacing for each wave, we can get much accurate results. Also, the number of points for each wave increases for upwind or BTD comparing to AB and CN for the same required accuracy because of the less accurate method. In practical problems with inflow turbulence, there is a wide range of frequencies or wavelength in the wind as we discussed in Chapter 2. If so, what is the minimum number of grid points one need to use to transport the smallest wave length or highest frequency of the wave? From numerical experiments, one can evaluate that. In practice, we will use minimum 4 spacing or 5 points as suggested by Ferziger and Peric (2002). We will discuss further in Chapter 5 when we apply CFD for flow over building.

3.4.3 Information Needed to Solve Unsteady Problems

For unsteady problems, the two major information needed are BCs and initial conditions. If proper initial conditions are not taken, the solution of nonlinear equation like NS equations may diverge to start with. If one can have a velocity field that satisfies continuity equation as an initial condition, the solution can converge much faster. Many situations, one may take potential flow solution as a start. The different BCs we discussed so far:

1) **Dirichlet** BCs. Here the unknown like velocity will be specified for the NS equations.
2) **Neumann** BCs. Here the derivative of the unknown will be specified. This means in those points; **the unknown has to be solved**.
3) Convective BC. This is specified at the outlet for the convective equation or the momentum equations. See Chapter 5 for further detail.

For the NS equations since one has velocities and pressure as unknowns, the BC get more complicated. The following BC's are the common ones:

1) Inlet BC. Here the velocities are specified but the pressure may be solved.
2) Symmetric BC. Here normal velocity on a surface is zero, and the normal derivative of the other velocities is zero. Here pressure needs to be solved.
3) Outlet BC. Here convective BC for velocities and for pressure either p specified or normal derivative of p specified.
4) On the wall no slip condition specified. This means velocities are zero and normal derivative of pressure is zero. In addition, for high Re flows or turbulent flows, one may also use law of the wall BCs. The law of the wall boundary condition is needed to improve the accuracy in approximating the logarithmic profile of the boundary layer.
5) For pressure solution where ever velocities are specified, normal derivative of p has to be specified. To have convergence, at least one-point p has to be specified. This has to be done for all solvers except SOR solver implemented by Hirt and Cook (1972) in SOLA method or TDMA line solver implemented by Patankar (1980).

Need for Law of the Wall Boundary Condition

In the atmospheric boundary layer, the velocity profile is represented as $U(z) = C_1 \ln[(z + z_0)/z_0]$ where z_0 is the roughness of the ground. The detail of the logarithmic profile will be discussed in Chapter 4. In the U momentum equation when the diffusion term d^2U/dz^2 is calculated at the point next to the wall, very high error occurs. For that let us consider $C_1 = 0.217$ and $z_0 = 0.01$ for the log profile and a grid spacing $h = 0.25$. Then U and its derivatives can be calculated analytically at $z = 0.25$ as follows:

$$U(z) = C_1 \ln \left[(z + z_0)/z_0\right] = 0.217 \ln (0.26/0.01) = 0.707$$

$$dU/dz = C_1/(z + z_0) = 0.217/0.26 = 0.868$$

$$d^2U/dz^2 = -C_1/(z + z_0)^2 = -0.217/0.26^2 = -3.21$$

FDM calculations at point 2:

Table 3.1 Comparison of exact derivatives to FDM derivatives and by law of the wall condition.

Point	1	2	3	Error in % @2
	z	0	0.25	0.5
	U(z)	0	0.707	0.853
	FDM-d^2U/dz^2 @2	-8.976	178%	
	Law of the wall @2	-4.09	27.4%	
	Exact @ 2			-3.21

$$dU/dz@\text{point } 2 = (U_3 - U_1)/2h = 0.853/0.5 = 1.706$$

and hence the error is $100(1.706 - 0.868)/0.868 = 96.5\%$

$$d^2U/dz^2@\text{point } 2 = (U_3 - 2U_2 + U_1)/h^2 = (0.853 - 2 \times 0.707 + 0)/0.125$$
$$= -8.976 \text{ and hence the error is } 100(8.976 - 3.21)/3.21 = 178\%$$

If we compare the exact values to FDM values at the point next to the wall (point 2) as shown in Table 3.1, we can see that there is very high error.

Instead, if the derivative d^2U/dz^2 is calculated at point 2 using CVM at faces e and w as shown later, the error is much less:

Wall

```
1         2            3
|-------h---------|--------h--------|
W------w-------P-------e--------E
       |--------h--------|
```

Using CVM: $d^2U/dz^2 @ 2 = [(dU/dz)_e - (dU/dz)_w]/h$

Here $(dU/dz)e = (U3 - U2)/h = (0.853 - 0.707)/0.25 = 0.584$

and $(dU/dz)_w$ as log profile at z=0.125 one get $(dU/dz)_w = 0.217/(0.125 + 0.01) = 1.607$

Hence,

$$d^2U/dz^2 = [(dU/dz)_e - (dU/dz)_w]/h = (0.584 - 1.607)/0.25 = -4.09$$

Error using the law of the wall condition: $100(3.21 - 4.09)/3.21 = 27.4\%$

So, at the wall, the law of the wall condition is implemented in computer programs as explained earlier.

3.5 Solution of the Incompressible Navier–Stokes (NS) Equations

After investigating the diffusion and convection phenomena numerically, here we will discuss the solution of the incompressible NS equations. The NS equations are nonlinear, and there are several different **segregated or sequential solutions** are available because the simultaneous solution of all the variables or **coupled solver** is time and storage consuming. Here coupled solver means all the variables (velocities and pressure) are solved simultaneously and sequential solver means when U velocity is solved from x-momentum equation, the other velocities V and W are considered known values and kept previous iterated value. Because of nonlinear equation, at each step, velocities and pressure are assumed to some value as a start to formulate the simultaneous equations as follows:

$$A^n\left(U_i^n\right)x^{n+1} = b$$

Here x^{n+1} is the unknown velocities and pressure at iteration number $n+1$. U_i^n is the velocity at iteration number, n and b are the known values at iteration number n. Here, i = 1 to 3 for velocities in the x, y, and z directions. For a coupled solution, the A matrix depends upon all the velocities. Initially, we do not know the velocities, and we take the previous time step or take some assumed value. Hence, A matrix depends upon the unknown U_i^{n+1} itself. After solving for U_i, the A matrix is updated, and the iteration has to be continued until convergence. This iteration is called **outer iteration**. To solve the A matrix iterative solver has to be used, and this iteration is called **inner iteration**. Because A matrix is updated at each time step and at each outer iteration, sequential solver is preferred. The steps involved in the **sequential solver** are as follows:

1) ISUB = 0.
2) Start outer iteration or subiteration: ISUB = ISUB + 1.
3) Solve for U by iterative method (inner iteration). Consider V, W, and P from previous iteration. Calculate the RNORM-U at the beginning of inner iteration. Same way, solve for V and W and calculate RNORM-V and RNORM-W.
4) Solve for P and calculate the RNORM-P or the continuity equation error. Update the velocities to satisfy continuity equation.
5) Check error < tolerance. If not go to step 2 until convergence or if ISUB < MISUB (max. ISUB). The step 2 to step 5 is called outer iteration.

In the preceding procedure, because of nonlinearity of the momentum equations, under relaxation factors are used in step 3. For details of implementation, one can refer to Ferziger and Peric (2002) and Patankar (1980). In step 4, if consistent pressure Poisson equation is used as in staggered grid, outer iteration convergence occurs very fast. For detail refer Selvam (2020) CFD notes. If not, several more number of iterations are needed for convergence. For nonstaggered grid one may have to use several outer iterations. For further discussions on pressure solution step and type of storage system is presented in Section 3.6.

In the sequential solver, the fractional step and SIMPLE solution procedure are the popular ones. The SIMPLE solver discussed in Patankar (1980) can be used for **steady and unsteady** NS equations. The fractional step procedure can be used only for **unsteady** NS equations, and the details are provided in Selvam (1997, 2020). The steady NS equations are used only in few situations. Hence, here unsteady solver will be used for all the applications. The steady NS equation solver using SIMPLE takes less computer time but it cannot capture the unsteady nature of the problem. For the steady SIMPLE solution, upwind scheme has to be used for the convection terms to get diagonal dominance, and the upwind scheme dampens the flow due to diffusive errors. The second-order methods such as CD, BTD, and QUICK are preferable for approximating the convection terms.

The following are the fractional step procedure for a staggered grid system:

1) Solve for Ui* from each momentum equation using the old pressure value.
2) Then solve for incremental pressure p′ to satisfy continuity equation from: $\nabla^2 p' = \nabla \cdot V^*$, where $V^* = iU_1 + jU_2 + kU_3$. This equation is formed by differentiating $U_i = U_i{}^* - dt\,(dp'/dx_i)$ in the corresponding direction and equating the $\partial U_i/\partial x_i = 0.0$.
3) Correct $U_i = U_i{}^* - dt(dp'/dx_i)$ and update the pressure as $P = P^* + p'$ by solving the following equations:

For nonstaggered grid system, one can refer to Selvam (1997).

3.6 Storage of Variables in Staggered and Nonstaggered Grid Systems

The velocities and pressure are located at different places in a staggered grid system, as shown in Figure 3.13. In the initial development of CFD, this is the method used because when coupled solvers are used, they found that A matrix is not positive definite when non-staggered grid system is used. To rectify the problem, staggered grid systems were used. This issue is called Babuska and Brezzi condition. For detailed discussions, one should refer finite element book like Bathe (1996). The benefit of staggered grid system is the pressure solver converges faster. The challenges are to apply in nonorthogonal grid systems. In the nonstaggered grid system, the variables are stored at the same point, as shown in Figure 3.13. This can be applied to nonorthogonal grid easily, as discussed by Ferziger and Peric (2002). In some situations, one faces convergence issues. As an example, for flow in a tornado chamber when nonstaggered grid system was tried the flow did not converge unless pressure is specified at a point as reported in Verma and Selvam (2020). On the other hand, when they used staggered grid system, they could get a converged solution.

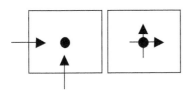

Figure 3.13 Staggered and nonstaggered storage system for cell-centered grid configuration.

3.7 Node and Cell-Centered Storage Locations

The variables can be stored at the nodes or cell-centered, as shown in Figure 3.14. Most of the popular commercial codes uses cell-centered storage. Selvam and Peng (1998) showed that building pressure is more accurate using node centered grid.

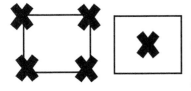

Figure 3.14 Storage of the variables at the nodes and cell centered.

3.8 Structured and Unstructured Grid Systems

In the structured grid, the variable T is stored as 2D variable. Here all the interior points P are connected to either 4 points for 2D or 6 points for 3D surrounding

Figure 3.15 Unstructured grid.

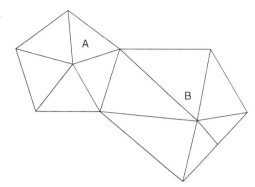

it in an orthogonal grid system. It is also called structured IJ or IJK system of storage. The same IJ or IJK system can also be stored in a 1D array, and some say this may increase the computational efficiency.

For an unstructured grid, a simple example with triangular region is shown in Figure 3.15. In the unstructured grid system used in FEM or CVM, the variables have to be stored as 1D variable and the coefficients needs to be stored either banded or skyline matrix form for direct solution or compact storage scheme for iterative solution. The compact storage scheme is well explained in Jennings (1977) and Saad (1996). This storage scheme is used by Selvam (1997) for their FEM solvers using unstructured grid.

3.9 Variable Storage Methods

To illustrate the storage methods of structured and unstructured grid we will consider a structured grid of 4x4 mesh which are spaced at equal distance of h = 1, as shown in Figure 3.16. The same grid is represented by structured and unstructured storage systems. The node number in the structured grid is identified by IJ system, as shown in Figure 3.16a, and the unstructured grid numbering has only one single number for each node, as shown in Figure 3.16b. In addition, the element or cell numbers are shown in Figure 3.16b at the middle of each element or cell.

In the structured grid, the unknown neighbors are identified for a IJ point is X(I,J), X(I+1,J), X(I−1,J), X(I,J+1), X(I,J−1) etc., where as in the unstructured grid system the neighbors are identified by element nodal connectivity matrix. The variables are stored in the unstructured grid system as (X(I),I = 1,NN), where NN is the number of nodes. Hence to solve the variables by iterative method, it is much easier to solve by structured grid system if the problem can be discretized in structured grid system. In the unstructured grid the equations have to be formulated in a matrix from Ax = b and each equation has to be multiplied by the number of variables connected to each node. Even though each node is connected to eight neighbors for all the interior points in this example, this may not be the case for any general unstructured grid. Each equation location in the A matrix is shown in Figure 3.17. For this, the CVM type 5 point stencil is shown for illustration. The location of east (e), west (w), north (n), and south (s) and point P is shown in Figure 3.17. If one

(a)

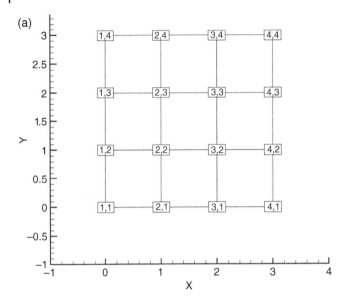

(b)

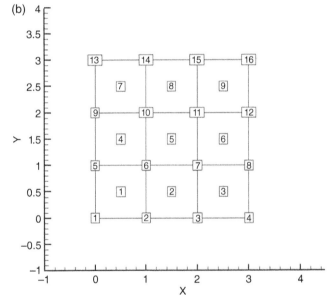

Figure 3.16 (a) Structured grid numbering. (b) Unstructured grid numbering.

needs to perform AX matrix operation in each method, let us assume that in the structured grid the A matrix coefficients are stored as $A(I,J,1) = AP$, $A(I,J,2) = -AE$, $A(I,J,3) = -AW$, $A(I,J,4) = -AN$, and $A(I,J,5) = -AS$ and for the unstructured grid the coefficients are stored as $(A(I,J), I = 1,NN$ and $J = 1,NN)$ as in Figure 3.17.

Node& Equation #	1	2	3	4	5	6	7	8	9	10	11	12	13	14	15	16
1	p	e			n											
2	w	p	e			n										
3		w	p	e			n									
4			w	p				n								
5	s				p	e			n							
6		s			w	p	e			n						
7			s			w	p	e			n					
8				s			w	p				n				
9					s				p	e			n			
10						s			w	p	e			n		
11							s			w	p	e			n	
12								s			w	p				n
13									s				p	e		
14										s			w	p	e	
15											s			w	p	e
16												s			w	p

Figure 3.17   Storage position for the A matrix in the unstructured grid system.

Let us say we store the AX in R. Then the algorithm for each method is as follows:
Structured grid:

```
DO J=1,IM
DO I=1,IM
R(I,J)=A(I,J,1)*X(I,J)+A(I,J,2)*X(I+1,J)+A(I,J,3)*X(I-1,J)+A(I,
J,4)*X(I,J+1)+A(I,J,5)*X(I,J-1)
END DO
END DO
```

Unstructured grid:

```
DO I=1,NN
SUM=0
DO J=1,NN
SUM=A(I,J)*X(J)
END DO
R(I)=SUM
END DO
```

For unstructured grid the storage method shown will not be used for computational efficiency. This is merely to show the method. In practice, the compact storage scheme will be used as discussed in the previous section.

3.10   Practical Comments for Solving the NS equation

1) For approximating the convection term, upwind procedure and central difference method are used. The upwind procedure may be preferred for flows using Reynolds average method of turbulence. For LES, central difference method is preferred. The upwind methods are diffusive and cannot transport turbulence accurately. But, the upwind method is introduced by Patankar and his group (1980) for solving steady NS equation as well as for unsteady NS equation with explicit methods. When CD method is used, the time step restriction is very severe for explicit methods. For implicit methods like Crank-Nicolson, one can use the CD method. The reason we prefer implicit method than explicit method for convection is the error in the momentum equation is reduced to the level of required accuracy by subiteration. Whereas by explicit method this cannot be done. For more numerical detail, one can refer to Selvam (2020) class notes and Ferziger and Peric (2002) book.
2) If the domain is orthogonal, then one can use line iteration for momentum equation and PCG method for pressure equation.
3) Staggered grid can solve the pressure equation more efficiently and without specifying the pressure at one point. Whereas in nonstaggered grid one has to specify the pressure at one point for faster convergence. For implementing BCs, the nonstaggered grid is preferred. Also, staggered grid is applicable for orthogonal grid, but for nonorthogonal grid nonstaggered grid is preferred.
4) In the NS equation, the velocity and pressure are coupled. Very rarely coupled solver is used. In the coupled solver, the velocities and pressure are solved simultaneously. Since the incompressible equation provides an unsymmetric matrix on the left-hand side, finding the proper solver is a major issue. Hence, in most of the practical problems, only uncoupled or sequential procedure is used. In the sequential procedure, common methods used are fractional step and SIMPLE.

3.11   Chapter Outcome

1) Introduction to the FDM for numerical modeling.
2) Computer modeling using diffusion-type equation. Exposure to direct and iterative solution. Gauss–Siedel introduced successive over relaxation methods.
3) Computer modeling of 1D wave equation to introduce convection modeling. Discussed explicit and implicit method of approximation for time-dependent problems. Challenges in convection modeling are discussed.
4) Central difference and upwinding concept for convection equation is introduced.
5) Stability conditions for explicit solution of convection and diffusion are discussed.
6) Solution of NS equation by sequential solution (fractional step and SIMPLE).
7) Staggered and nonstaggered grid system, cell centered and node centered storages, and structured and unstructured grid systems are also discussed.
8) Visualization using tecplot is introduced using diffusion problem.

Problems

1 For the unsymmetric chamber flow described in P#2, using 4×4 mesh perform hand calculations and get the final stream function values. Solve by the direct method like Gaussian elimination and the iterative method of Gauss–Siedel. Perform iteration with minimum five iterations.

2 Prepare input for the flow in a chamber with flow from the bottom boundary and flow goes out from the top right boundary problem. The domain size is 3×3. Use a grid of size 49×49. Plot the contour and vector plot using tecplot or any other visualization software. The details of modeling are described later:

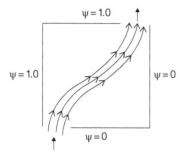

3 For the flow in a chamber problem as in problem 2, determine the optimum RF. Also, plot RF vs. number of iterations for 49×49 grid.

4 Consider a log flow of the wind $[u(z) = 0.217 \ln((z + 0.01)/0.01), v = 0, w = 0]$ and calculate the vorticity $[du/dz]$ by FDM and compare with closed form $[du/dz = 0.217/(z + 0.01)]$. Consider 5 points from the ground with a spacing of $h = 0.25$ and calculate the vorticity or du/dz for the interior 3 points by central difference. At $z = 0$, $u(0) = 0.0$ is used. Report the error in percentage $= 100 \times (\text{exact-FDM})/\text{exact}$. Calculate the error for different grid spacing say $h = 0.1$. Give details.

5 Vary the IM, CFL, and TTIME in the programs **w1dcn.f** and **w1dab.f** and report the performance. Keeping XL $= 2$ units (i) for a given time and CFL, if IM is varied, the minimum number of points/per wave and for high accuracy what is the number of points/wave can be determined. Using 6 points/wave and 12 points/wave study the performance. Keep CFL $= 0.5$; (ii) variation of CFL shows increased accuracy of transport with less CFL. For 12 points/wave, study the effect of CFL by varying it from 0.1 to 0.9; and (iii) not much difference if TTIME is increased for this problem if TTIME > 2 units.

6 Compute the storage requirements to solve the Poisson problem in a 3D domain with $100 \times 100 \times 100$ mesh using the following methods:
   a) Using CVM with 7 point stencil for interior points.
   b) Direct solution by storing the full A matrix.
   c) Using banded solution procedure and assuming that the semi bandwidth may be 10 000. Assume also that the matrix is symmetric.

References

Bathe, K.J. (1996). *Finite Element Procedures*. Englewood Cliffs, NJ: Prentice Hall.

Biringen, S. and Chow, C.Y. (2011). *An Introduction to Computational Fluid Mechanics by Example*. John Wiley & Sons, Inc.

Canuto, C. (1988). *Spectral Methods in Fluid Dynamics*. New York: Springer-Verlag.

Crandall, S.H. (1956). *Engineering Analysis: A Survey of Numerical Procedures*. New York: McGraw-Hill.

Ferziger, J.H. and Peric, M. (2002). *Computational Methods for Fluid Dynamics*, 3ee. Springer.

Hirt, C.W. and Cook, J.L. (1972). The calculation of three-dimensional flows around structures and over rough terrain. *Journal of Computational Physics* 10: 324–340.

Jennings, A. (1977). *Matrix Computation for Engineers and Scientists*. London: Wiley.

Mansouri, Z., Selvam, R.P., Chowdhury, A. and (2020). Grid spacing effect on peak pressure computation on the TTU building using synthetic inflow turbulence generator. Report, Department of Civil Engineering, University of Arkansas.

Orlandi, P. (2000). *Fluid Flow Phenomena: A Numerical Toolkit*. Kluwer Academic Publishers.

Patankar, S.V. (1980). *Numerical Heat Transfer and Fluid Flow*. Taylor & Francis.

Saad, Y. (1996). *Iterative Methods for Sparse Linear Systems*. Boston: PWS Pub. Co.

Sarkar, S. and Selvam, R.P. (2009). Direct numerical simulation of heat transfer in spray cooling through 3D multiphase flow modeling using parallel computing. *Journal of Heat Transfer* 131: 121007-1–121007-8.

Selvam, R.P. (1996). Computation of flow around Texas Tech Building using k–ε and Kato-Launder k–ε Turbulence Model. *Engineering Structures* 18: 856–860.

Selvam, R.P. (1997). Finite element modelling of flow around circular cylinder using LES. *Journal of Wind Engineering and Industrial Aerodynamics* 67 & 68: 129–139.

Selvam, R.P. (1998). Computational procedures in grid based computational bridge aerodynamics. In: *Bridge Aerodynamics* (ed. A. Larsen and S. Esdahl), 327–336. Rotterdam: Balkema.

Selvam, R.P. (2020). Introduction to computational fluid dynamics and heat transfer. CFD class notes, Report, Department of Civil Engineering, University of Arkansas.

Selvam, R.P. and Peng, Y. (1998). Issues in computing pressure around buildings. In: *Structural Engineering World Wide 1998* (ed. N.K. Srivastava), 941pp. New York: Elsevier Paper reference T171-3.

Tannehill, J.C., Anderson, A.D., and Pletcher, R.H. (1997). Different explicit and implicit methods of approximating the convection and diffusion equations are presented well. In: *Computational Fluid Mechanics and Heat Transfer*, 2ee. Philadelphia: Taylor & Francis.

Verma, S. and Selvam, R.P. (2020). CFD to VorTECH pressure-field comparison and roughness effect on flow. *Journal of Structural Engineering* 146 (9): 04020187-1–04020187-12.

4

Introduction to Wind Engineering

Wind Effects on Structures and Wind Loading

To understand the effect of wind on buildings, bridges, and other structures, one needs to understand the characteristics of wind. The wind is a turbulent flow as discussed in Chapter 2. When the wind interacts with structures depending on the structural property of the structure, different wind phenomenon happens. This can be simply explained by watching a tree interacting with wind. In a tree, the trunk is very stiff and the period of the structure is small comparing to the branches and leaves. The leaves are far more flexible than branches. Due to reasonable speed of wind, the trunk seems not moving much comparing to branches and leaves. The oscillation of branches can be seen more clearly than trunk. The leaves oscillate far more than branches and we may describe that oscillation as fluttering of the leaves. Hence, to understand these complex phenomena, one needs to have strong background in fluid mechanics, turbulence, wind engineering, structural dynamics, and random vibration. In this chapter, some introduction to wind engineering is provided. For more detailed discussions, one should refer to Dyrbye and Hansen (1999), Holmes (2007), Liu (1991), and Simiu and Scanlan (1986). In this text, no discussion is made on introduction to structural dynamics and one should take at least one course on structural dynamics. For a simple introduction to structural dynamics, one can refer to Selvam (2017b).

The wind that interacts with structures can be classified into three main categories:

1) Straight-line (SL) boundary layer wind – this also includes hurricane-type wind;
2) Thunderstorm (TS) downdraft wind;
3) Tornado.

The wind profile variations from the ground for the SL and TS are shown in Figure 4.1. The tornado wind profile from the ground can be assumed to be similar to SL wind and only thing is the wind direction changes all around from the center of the tornado. In all cases on the ground, the wind velocity is zero due to the presence of viscosity in the air. More details on TS wind profile are available from Selvam and Holmes (1992) and Holmes (2007), and more details on tornado wind profile are available from Selvam and Millett (2003) and Strasser and Selvam (2015).

Of the three types, ASCE 7-16 considers only SL boundary layer wind at this time. There is some understanding on the other two types, but they are not to the level to get it in the

Computational Fluid Dynamics for Wind Engineering, First Edition. R. Panneer Selvam.
© 2022 John Wiley & Sons Ltd. Published 2022 by John Wiley & Sons Ltd.

(a)   (b)   (c)

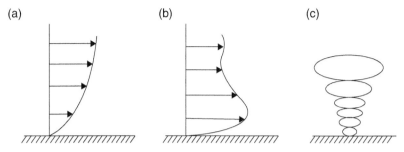

Figure 4.1   Wind profile for (a) straight-line boundary layer wind, (b) thunderstorm downdraft wind, and (c) tornado.

code as mentioned in ASCE 7-16; see Section 26.14. A brief introduction to wind engineering is attempted in this chapter.

4.1   Wind Velocity Profile Due to Ground Roughness and Height

The wind velocity changes from zero to a maximum at a certain height from the ground. The maximum velocity reaches at different height depending on the ground roughness. This can be illustrated in Figure 4.2 for various ground roughness or exposure as discussed in Davenport (1961). The code considers this effect by classifying the ground roughness as Exposure A, B, C, and D. This is also mentioned in Figure 4.2. In the middle of the downtown of a big city (Exposure A), the velocity varies from the ground very slowly comparing to velocity variation on a water surface like sea (Exposure D). The velocity variation in the midst of medium size houses in reasonable town (Exposure B) has little faster variation compared to Exposure A, and the variation on a grass surface with small tress (Exposure C) is much slower than Exposure D and faster than Exposure B. The ASCE incorporates these variations in the wind load calculations, and they also illustrate the ground roughness

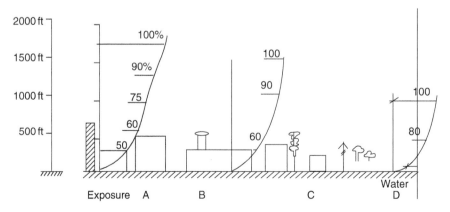

Figure 4.2   Mean velocity profile over different ground roughness.

with aerial pictures in the commentary with some description in Table C26.7-2. At this time, the code does not provide wind profile for Exposure A. They assume that for that situation, one needs special attention and the interested part should contact wind engineers.

Mean Wind Speed: The wind speed reported in the ASCE 7-16 for various parts of the United States is 3-s gust wind speed for Exposure C. It is measured at 10 m from the ground level. This wind speed also includes the hurricane wind. The 3-s gust wind speed means the maximum mean wind in three seconds (U_3). The other methods of reporting wind speed are fastest mile, one-minute, and mean hourly.

Fastest Mile: The fastest mean wind speed is recorded by the anemometer for 1 mile of distance (Ufm). This means if the mean wind speed is higher, the time to travel 1 mile will be less.

One-minute: Maximum mean wind speed in one minute (U_{60})

Mean Hourly: Maximum mean wind speed over one hour (U_{3600})

Different countries use different averaging time for wind speed. Eurocode uses 10 minute (U_{600}) averaging time and Canada uses 1 hour averaging time (U_{3600}). The longer the averaging time, the lesser the mean wind speed. To convert from one system to another, Durst curve provided in the ASCE commentary can be used, for example, $U_3/U_{3600} = 1.5$ from Durst curve. If we know one, we can calculate the other.

4.1.1 Wind Velocity with Height

The wind velocity varies with height as shown in Figure 4.2 for different ground roughness. The velocity profile with height can be calculated by two different methods. The meteorological world uses log-law and ASCE 7-16 uses power-law. The effect of the height is considered by ASCE through the term K_z.

Using the log-law, the velocity at any height z from the ground is calculated as follows:

$$U(z) = (u^*/\kappa) \ln\left[(z + z_0)/z_0\right] \tag{4.1}$$

where u^* is the frictional velocity, κ is the Von Karman constant equal to 0.4, and z_0 is the roughness of the ground in meters or feet.

Same way the power-law profile is calculated as:

$$U(z) = U_1(z/z_1)^\alpha \tag{4.2}$$

where U_1 is the wind speed at height z_1 and α is the power-law exponent depends on the terrain roughness. The value of α for different terrain is given in ASCE 7-16, as shown in Table 26.11-1. The z_0 for different terrain is shown in Table 4.1 (taken from Stull 2017).

In the computational fluid dynamics (CFD) applications, the log-law profile is preferred in many occasions. As an example, if one uses k-ε turbulence model with the log-law profile and the parameters given in equation later for k and ε at the inflow, the governing equations are satisfied for channel flow:

$$k = u^{*2}/\sqrt{Cr} \quad \varepsilon = u^{*3}/[\kappa(z + z_0)]$$

Table 4.1 Roughness length z_0 in meters for different terrain.

z_0 (m)	Classification	C_D	Landscape
0.0002	Sea	0.0014	Sea, paved areas, snow-covered flat plain, tide flat, smooth desert
0.005	Sooth	0.0028	Beaches, pack ice, morass, snow-covered fields
0.03	Open	0.0047	Grass prairie or farm fields, tundra, airports, heather
0.1	Roughly open	0.0075	Cultivated area with low crops and occasional obstacles (single bushes)
0.25	Rough	0.012	High crops, crops of varied height, scattered obstacles such as trees or hedgerows, vineyards
0.5	Very rough	0.018	Mixed farm fields and forest clumps, orchards, scattered buildings
1.0	Closed	0.030	Regular coverage with large-sized obstacles with open spaces roughly equal to obstacle heights, suburban houses, villages, mature forests
≥ 2	Chaotic	≥ 0.062	Centers of large towns and cities, irregular forests with scattered clearings

Source: Taken from Stull (2017).

The preceding parameters are derived and used by Selvam (1990). If the power-law profile is used, it may not satisfy the governing equations and hence at some distance from the inflow the profile may change due to this issue. Similarly, if one uses simple mixing length theory as discussed in Schlichting (1968) to calculate the turbulent eddy viscosity in the boundary layer as:

$$\nu_t = \kappa(z + z_0)u^*$$

then this expression along with log-law profile will satisfy the governing equation but not the power-law profile.

Example 4.1 Let us consider an open terrain (Exposure C) with $\alpha = 1/7 = 0.143$ and $z_0 = 0.035$ m. The velocity at height 10 m from the ground is 10 m/s. Find the velocity at $z = 20$ m.

For log-law, we need to find the u^* as follows:

$$U(10) = (u^*/0.4)\ln(10.035/0.035) = 10, \text{ then } u^* = 10^*0.4/\ln(10.035/0.035) = 0.707$$

Hence, $U(20) = (0.707/0.4)\ln(20.035/0.035) = 11.22$ m/s

For power-law, $U(20) = 10(20/10)^{0.143} = 11.041$ m/s

4.2 Topographic Effect on Wind Speed

When the wind flows over a hill or a valley, the wind profile close to the ground changes. In the case of wind flow over hills, there are situations where in the wind speed close to the hill top say 10 m from the ground increased by 80% more than the normal boundary layer wind

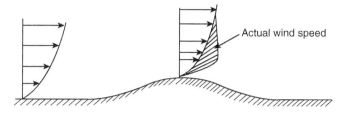

Figure 4.3   Speed-up due to wind flow over a hill.

for a slope of 0.25 as reported by Selvam and Smith (1989). The wind speed change due to hill is very much close to the ground as shown in Figure 4.3. Hence, the code also considers the effect through a factor K_{zt}.

4.3   Wind Speed and Wind Pressure

The Bernoulli equation relates the pressure to wind speed by $p = C\rho V^2$ where C is a pressure coefficient. The same approach is taken by ASCE to relate pressure to wind speed. The p is related to velocity in mph by $0.00256\,V^2$. Here, the constant takes care of the 0.5 (density) (conversion from mph to ft/s). The detail of the conversion is provided in ASCE C26.10.2. In addition, to consider directional, roughness, and terrain effect, K_d, K_z, and K_{zt} terms are introduced and the velocity pressure is calculated by ASCE Equation 26.10-1 as follows:

$$q_z = 0.00256 K_d K_z K_{zt} V^2 \qquad (4.3)$$

where V is the velocity in mph at 30 ft height from the ground obtained from the map, K_d is the directional factor, K_z is the type of ground roughness or exposure, K_{zt} is the topographic effect, and q_z is in psf.

   The wind pressure does not have on its own any effect on the structure. This is related to different structures by different relationships, and we will see them in the coming sections.

4.4   Wind and Structure Interaction

Even within the SL boundary layer-type wind, when the wind interacts with structures and surroundings, several phenomena happen. The fluid–structure interaction (FSI) is also called aerodynamic effect.

   Overall, the aerodynamic effect of wind can create dynamic response on a structure in the following three ways as reported by Davenport (1995) and Dyrbye and Hansen (1999):

$$F = F_t + F_s + F_{ad}$$

where F_t is the force induced by turbulent fluctuations and produces background and resonant responses in the along-wind and cross-wind directions, F_s is the force caused due to vortex shedding and mostly influences the cross-wind direction, F_{ad} is the force created by motion of the structure. This is mainly due to aerodynamic damping and it controls the resonant response amplitude. Negative aerodynamic damping is mostly associated with crosswind motion. In extreme cases this led to aerodynamic instability. The clear understanding of the contribution of the aerodynamic damping during vortex shedding is not well understood. Selvam et al. (2001, 2002) applied CFD modeling to compute the critical velocity for flutter. There the effect of negative aerodynamic damping is clearly illustrated. Further discussions are made in Section 4.4.3 and Chapter 6.

4.4.1 Shape Effect

Depending on the shape of the structure, the force on the structure along the wind direction can change. The effect can be considered using the force coefficient C_f. The force is related to C_f as follows:

$$F = AC_f \rho V^2 / 2 \tag{4.4}$$

where A is the cross section perpendicular to the flow or the projected area. As an example, the C_f for a square, circular, and elliptical (L/D = 8) cross sections are 2.2, 1.2, and 0.25, respectively, as reported in Sadraey (2009). Similarly, for a smooth bridge cross section, C_f is reported to be around 0.5 in Selvam et al. (2002). Hence, if the shape of the structure is aerodynamic as in airplane wing or bridge, the C_f will be small. On the other hand, if the body is bluff as in square or circular, the C_f is higher. The force that is developed along the wind direction is also called along-wind force. Yousef et al. (2018) calculated the forces on a cubic building and a dome for straight and tornadic wind using CFD. They also considered the ASCE 7 loading for comparison. From their comparisons, one can see that the prism force coefficients on the roof are more than three times the dome for SL wind and for tornadic wind around two times more. In the same way, the ASCE 7 reports different pressure coefficients for flat ($\theta < 7°$), gable, and hip ($7° < \theta < 20°$) roof. A 3D view of gable and hip roof is shown in Figure 4.4. The hip roof is far more aerodynamic shape than gable roof. Because of that, the peak pressure coefficients (components and cladding) on the edges are varying from -2.3 to -3.2 for flat roof, -3 to -3.6 for gable roof, and -2.4 to -2.6

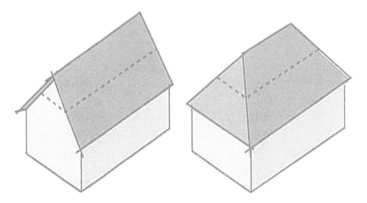

Figure 4.4 Gable and hip roof.

for hip roof, respectively. This is the reason in the hurricane prone area, flat or hip roof is recommended comparing to gable roof. Gable roof is preferred in the high snow area, where the snow can fall down easily.

4.4.2 Structural Dynamic Effect in the Along-Wind Direction

The wind velocity changes on its own due to turbulence in the wind. This change in wind speed with time interacts with structure, and it enhances the force along the wind direction depending on the period of the structure Ts. This effect is considered in the code using a coefficient called gust effect factor G. The code considers structures with a period less than 1 second or frequency greater than 1 Hz to be *rigid structure* and suggests a G value of 0.85. If the period is more than one second, then G value has to be calculated using an analytical function as discussed in ASCE 26.11.5. To use the equations, one has to have the Ts and damping coefficient for the structure. These structures are called *flexible* or dynamically sensitive structures. The effect of wind on rigid and flexible structures are is illustrated in Figure 4.5.

The amplitude due to dynamic effect in this case will be less comparing to total deflection or response, because this happens from the mean response as shown in Figure 4.5. Wind turbulence has several frequencies, and hence mostly statistical concepts are utilized to arrive at the gust factor. For details, one can refer to books like Dyrbye and Hansen

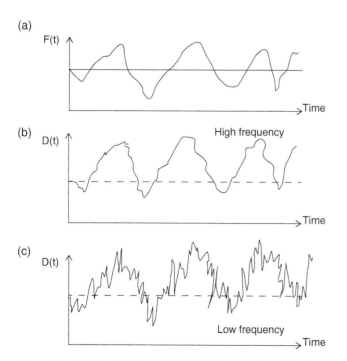

Figure 4.5 Effect of wind on rigid (high frequency) and flexible structure (low frequency): (a) reference wind speed with time, (b) displacement on rigid structure with time, and (c) displacement on flexible structure with time.

(1999) and Holmes (2007). This technique is first introduced in wind engineering by Davenport (1961). The along-wind force is also called as drag force when dealing with flow around airfoil and bridge. Example problems for rigid and flexible structure using ASCE 7 are illustrated in Selvam (2017b).

4.4.3 Structural Dynamic Effect in the Across-Wind Direction

When wind flow over any bluff body like circular cylinder, *vortex shedding* occurs behind the cylinder as shown in Figure 4.6. This vortex forms alternatively on the top and bottom of cylinder, and this produces forces perpendicular to the flow. This is also called lift force (C_l). This force is a time-varying force. As an example, the drag (C_d) and lift (C_l) force coefficients for a fixed circular cylinder at Reynolds number 100 are plotted in Figure 4.7. Here, C_d and C_l can be calculated from the force in x (F_x) and y (F_y) directions as follows:

$$C_d = F_x/(\rho U^2)/2 \quad \text{and} \quad C_l = F_y/(\rho U^2)/2$$

These coefficients are calculated using computer modeling, and for further details, one can refer to Strasser and Selvam (2015). From the plot, one can see that the time-varying lift force C_l in green will produce dynamic effect if the structure is wind-sensitive. Especially, flexible structures like tall buildings, chimneys, thin poles, signal posts, etc., will be affected by across-wind oscillations.

The drag and the lift force oscillation periods for a fixed cylinder are 2.5 and 5 time units, respectively, as shown in Figure 4.7. The shedding frequency is directly proportional to upstream velocity (U) and inversely proportional to the diameter of the cylinder (D). The relationship can be written as:

$$f_{vs} = S_t U/D \tag{4.5}$$

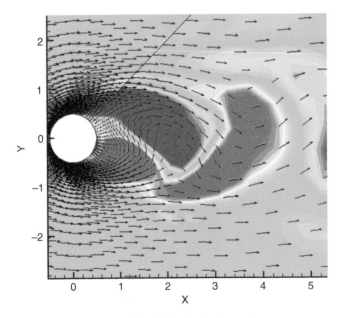

Figure 4.6 Vortex shedding behind a circular cylinder.

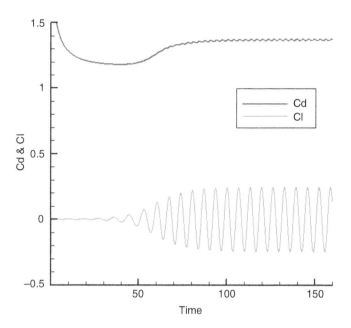

Figure 4.7 Time variation of drag (C_d) and lift (C_l) force coefficients.

where f_{vs} is the vortex shedding frequency and S_t is the Strouhal number. The flow velocity U is also related to the viscosity μ of the fluid by the Reynolds number Re as follows:

$$Re = \rho UD/\mu = UD/\nu \qquad (4.6)$$

where ρ is the density and ν is the kinematic viscosity of the fluid.

For a circular cylinder, the St remains a constant value of 0.2 in the subcritical range ($Re \leq 3 \times 10^5$) and 0.25 in the transcritical range ($Re \geq 3 \times 10^6$) and for a square cylinder St = 0.11 as reported by Buchholdt and Moossavi (2012). The amplitude of the C_l is 0.5 for circular cylinder in the subcritical range, 0.25 in the transcritical range, and 0.6 for square cylinder.

Looking into wind engineering literature (Blevins 1977; Holmes 2007; Sumer and Fredsøe 2006), one can say that the Strouhal number St for a flat plate is around 0.15–0.17. Holmes (2007) provides a range of St from circular cylinder to almost flat horizontal plate to be 0.2–0.08. Here, St = fU/D or U/(DT) where D is the distance perpendicular to the flow, f is the frequency, and T is the period of the shedding.

Due to this time-varying load on the structure, dynamic response will happen. Especially, if the structural frequency f_{st} is close to or equal to the f_{vs}, then resonance can occur. It has been observed that when f_{st}/f_{vs} is close to 1, the structure oscillates with structural frequency for a range of reduced velocity $U^* = U/(f_{st}D)$ from 4 to 6. This phenomenon is called *lock-in* condition. For further details, one can refer to Holmes (2007) and Dyrbye and Hansen (1999).

If f_{st} is known, one can find the critical velocity for resonance from Equation (4.7) as:

$$U_{crit} = f_{st}D/S_t \qquad (4.7)$$

The response of structure due to cross-wind response is very complicated and ASCE 7-16 does not consider the effect of it. Two major methods are discussed in Holmes (2007) and other codes follow any one of the procedures. To understand the issue, let us introduce a simple approach assuming complete correlations in the vertical direction. The dynamic load factor (DLF) for a harmonic loading with damping ξ for a single-degree freedom system is

$$\text{DLF} = y_{dy}/y_{st} = 1/(2\xi) \tag{4.8a}$$

$$Y_{dy} = (F/k)/(2\xi) = F/(2M\omega^2\xi) = C_1(\rho U^2 DH/2)/(2M\omega^2\xi)$$

Substituting Strouhal number St $= f_{st}D/V$, $\omega = 2\pi f_{st}$

$$Y_{dy} = (\rho_f HD^3/M)(C_1/[16\pi^2\xi S_t^2]) = (\rho_f D^3/m)(C_1/[16\pi^2\xi S_t^2]) \tag{4.8b}$$

where F is the amplitude of the lift force, k is the stiffness of the structure, D is the diameter of the cylinder, H is the height of the structure, M is the total mass of the structure, $m = M/H$ is the mass per unit length, ρ_f is the density of air, and C_1 is the lift coefficient. Here, the aerodynamic damping effect is not considered.

Example 4.2   Let us consider a steel circular mast of diameter 0.5 m. The thickness of the mast is 5 mm. The problem is similar to the one considered for CFD modeling in Belver et al. (2012). The frequency of the structure $f_{st} = 0.43$ Hz, Re $= 4 \times 10^4$, density of steel $\rho_s = 7772$ kg/m^3, mass per unit length of the mast m $= 60.43$ kg/m, density of air $\rho_f = 1.21$ kg/m^3, U $= 1.1$ m/s, and $\xi = 0.025$. Since it is subcritical range $C_1 = 0.5$ is taken:

$$S_t = f_{st}D/U = 0.43^*0.5/1.1 = 0.2$$

$$Y_{dy} = (\rho_f D^3/m)(C_1/[16\pi^2\xi S_t^2]) = (1.21^*0.5^3/60.43)^*0.5/(16^*\pi^{2*}0.025^*0.2^2) = 0.008 \text{ m}$$

Belver et al. (2012) reported a value of 0.006 m and the approximate computation is in the reasonable range. For more detailed analytical computation, one can refer to Holmes (2007).

The design procedure suggested by Griffin and Ramberg (1982) will be very useful for design application as reviewed by Selvam (2017a). Griffin and Ramburg report that the maximum displacement with respect to the cylinder diameter (Y_{max}/D) varies from 1 to 2 for lower reduced damping ratio of 0.01, and this ratio Y_{max}/D is 0 for reduced damping ratio of 5. Here, reduced damping ratio is defined as $8\pi S_t^2\xi/\rho D^2$. Selvam (2017a) also reports that galloping and flutter have large deformation due to negative damping comparing to the thickness of the structure, and the oscillation is closer to the period of the structure. The effect is more due to structural motion than vortex shedding. Whereas, the motion due to vortex shedding will have oscillation close to Strouhal frequency and the amplitude is higher when the resonance occurs.

Aerodynamic Damping: The aerodynamic damping can be mathematically explained by taking a single-degree freedom system with a following forces:

$$md^2y/dt^2 + cdy/dt + ky = f(t) = a + bdy/dt$$
$$md^2y/dt^2 + (c-b)dy/dt + ky = a$$

In the preceding equations, m, c, and k are the mass, damping coefficient, and stiffness of the spring–mass system, respectively. The right-hand side force "a" is a constant value and "b" is the coefficient of aerodynamic damping force. The solution to this equation as reported by Warburton (1976) is

$$y(t) = a/k + A \exp[(b - c)t/(2m)] \sin(\omega t + \alpha)$$

where A and α depend on the initial conditions. In the preceding equation, if $c > b$ or b is negative, then the system is similar to usual positive damping and the displacement amplitude will exponentially decreases if "a" is zero. On the other hand, if $b > c$, then the overall damping becomes negative and the amplitude of the vibration increases exponentially. This type of force is produced when the wind interacts with structure, and the negative damping is called aerodynamic damping. The study of FSI is called aeroelasticity. The negative damping can produce self-excited vibrations and finally leads to failure as in bridge flutter and galloping of transmission cable.

Galloping and Flutter Phenomena: Galloping is a vertical motion instability. A simple criterion is introduced by Den Hartog (1932) as a static calculation $C_d + dC_l/d\alpha < 0$. For this, one needs to plot the C_l at various angle of attack α as shown in Figure 4.8. This graph is taken from Parkinson and Brooks (1961). In the criterion, the drag coefficient C_d is always positive and when the slop of C_l is negative, then galloping can occur. In the figure, the slope changes to negative direction for the square around 15° and for the longer rectangle in the x-direction around 8°. There are two major ways one can calculate the critical velocity for galloping using CFD. One way is to use partial CFD and structural dynamics, and the other

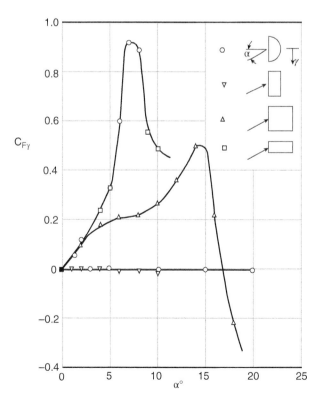

Figure 4.8 Lift versus angle of attack for square, rectangle, and half circle cylinders. *Source:* Taken from Parkinson and Brooks (1961).

way is fully as a FSI problem as described in detail by Selvam (2017a). Partial CFD, one does CFD calculation for the C_l at various angle of attack, and then the structural dynamics is done as illustrated in Selvam (2017). The procedure is reported in Parkinson and Smith (1964) for wind tunnel testing.

Similarly, vertical (heaving) and rotational (pitching) motion together can cause flutter instabilities in bridges. The famous example is the Tacoma narrow bridge failure. Similar to galloping, Simiu and Miyata (2006) say that when the slope of $dC_l/d\alpha$ and $dC_M/d\alpha$ becomes negative, it causes the flutter condition. Here, C_M is the moment coefficient.

As an example, when the wind interacts with bridges, beyond critical velocity for flutter, the oscillation increases continuously due to negative damping as shown in Figure 4.9b, and

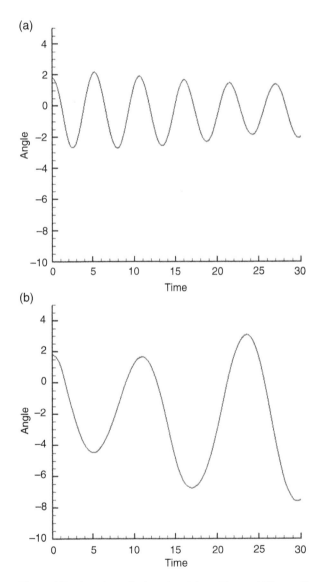

Figure 4.9 Aerodynamic damping: (a) positive and (b) negative.

if the velocity is less than the critical value, the oscillation dampens as shown in Figure 4.9a. The critical velocity for flutter is around 70 m/s for the Great Belt East bridge suspension span as reported in Selvam et al. (2002). Here, the structure is modeled as two degrees of freedom (pitching and heaving motion).

4.5 Opening in the Building

The opening in the building can change the pressure on the structure. If on one side of the building there is a big window and let us say it is opened, the opening will change the flow condition around there and the pressure inside the building and outside the building on a wall influences it simultaneously. This effect is considered in the code through a term GCpi and the effect is high when the opening area is large.

4.6 Phenomena not Considered by the ASCE 7-16

Several of the following phenomena are not considered at this time as mentioned in the ASCE7-10, 27.1.2:

1) Across-wind loading
2) Vortex shedding
3) Instability due to galloping or flutter
4) The wake of upwind obstruction
5) Channeling effect due to site location

The first two are considered some level in the preceding section. The other phenomena can be understood by referring to standard wind engineering text like Holmes (2007). The dynamic effect of wind is considered only for the along-wind direction for flexible building and the across-wind response is not considered. The response due to cross-wind can be calculated using other codes as discussed in Dyrbye and Hansen (1999).

The commentary of ASCE-7 provides the following guidelines to consider the above effect for buildings:

1) If the building height H is more than 400 ft.
2) If the $H/B_{min} > 4$, where B_{min} is the minimum width of the building.
3) If the lowest frequency of the building $n_1 < 0.25$ Hz or period $T_1 > 4$ s.
4) If the reduced velocity $U^* = U_z/(n_1 B_{min}) > 5$ where $z = 0.6H$ and U_z is the mean hourly velocity.

4.7 ASCE 7-16 on Method of Calculating Wind Load

The ASCE 7-10 provides methods to calculate the wind load for main wind force resisting systems (MWFRSs) and components and cladding (C&C). The MWFRS loads are overall loads on the structure. Since this covers larger area, the pressure coefficients will be lower than C&C pressures. The C&C refers to anything which is not MWFRS such as window panels, roof panels, etc. For design of window glass panels, roof fasteners for a specific area C&C section will be used. There one can see that if the area decreases, the pressure coefficient increases. In addition, the value of pressure also changes depending on the location in the building. Around the corners and edges of the building, the design pressure is high and especially the roof edge and corner values are higher than the wall edge and corner values. The pressure for the C&C is obtained from ASCE Chapter 30.

Different methods to get the wind–structure interaction effects using the ASCE 7-16 are analytical, wind tunnel, and computer model. For simple configurations for which the ASCE 7-16 is applicable, wind load can be obtained using the analytical methods. Under analytical methods, three major methods are provided:

1) Directional procedure for building of all heights (ASCE Chapter 27 – Method 1)
2) Envelope procedure for low-rise building (ASCE Chapter 28 – Method 2)
3) Directional procedure for other structures (ASCE Chapter 29 – Method 3)

The first method or Method 1 is general and applicable to any building heights which are regular-shaped. The forces may be lesser than Method 2 but may take more work. Here, the effects of wind flow on the windward and leeward side are considered as shown in Figure 4.10. When the wind flow over and around a building, the force created on the windward part of the building is compressive in nature on the wall. The code considers the compressive effect of the pressure to be positive. On the roof and on the leeward wall, the pressure is going to be negative: i.e. the forces will pull away from the wall. The positive and negative pressure effects on all these walls and roof are considered in Method 1. The variation of pressure on the roof is considered far more in detail than Method 2. The building can be rigid (frequency <1 Hz) or flexible (frequency >1 Hz) in Method 1 or Method 3, but Method 2 can only be applied to rigid buildings.

On the other hand, Method 2 is easy to apply, but there are restrictions to apply as discussed in ASCE Chapter 28. First of all, they have to be low-rise building and it is defined as building having height h less than 60 ft and h does not exceed the least lateral dimension of the building. Under Method 2, there are two methods called part 1 and part 2. Part 1 considers the pressure on the windward and leeward separately for any height of a building, and this has the provision to consider the effect of opening on the wall, parapets, and roof overhangs. Part 2 cannot consider the above-mentioned details. The part 2 is mainly for enclosed simple diaphragm building. In part 2, the windward and leeward pressure are considered together. The part 2 procedure is the easiest to apply. The Method 3 is applicable to

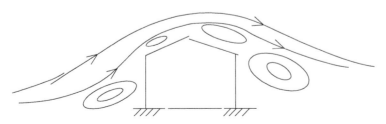

Figure 4.10   Wind flow over a building.

structures, such as roof top structures, solid signs, chimneys, open signs, and lattice frameworks. These structures are not considered in Method 1 or 2. For the application of ASCE 7-16 to calculate wind load on several different structures, one can refer to Mehta and Coulbourne (2013).

In the calculation of wind load using Method 2 of part 2, separate wind loads are provided for the interior and edge of the building. The edge loads are far higher than the interior part of the building. The edge width to be considered is provided in Figure 28.5-1 of ASCE. The formula to calculate the edge with:

$$2a = 2[\text{Min}(0.1b, 0.4h) > \text{Max}(0.04b, 3')]$$

where b is the lease lateral dimension and h is the mean roof height.

References

Belver, A.V., Rossi, R., and Iban, A.L. (2012). Lock-in and drag amplification effects in slender line-like structures through CFD. *Wind and Structures An International Journal* 15: 189–208.

Blevins, R.D. (1977). *Flow-induced Vibration*. New York: Van Nostrand Reinhold Co.

Buchholdt, H.A. and Moossavi, N.S.E. (2012). *Structural Dynamics for Engineers*. ICE Publishing.

Davenport A.G. (1961). A statistical approach to the treatment of wind loading on tall masts and suspension bridges. Ph.D. thesis, University of Bristol, Bristol, England.

Davenport, A.G. (1995). How can we simplify and generalize wind loads? *Joirnal of Wind Engineering and Industrial Aerodynamics* 54 (55): 657–669.

Den Hartog, J.P. (1932). Transmission line vibration due to sleet. *Transactions of AIEE* 51: 1074–1086.

Dyrbye, C. and Hansen, S.O. (1999). *Wind Loads on Structures*. New York: John Wiley & Sons.

Griffin, O.M. and Ramberg, S.E. (1982). Some recent studies of vortex shedding with application to marine tubulars and risers. *ASME. Journal of Energy Resources Technology* 104 (1): 2–13.

Holmes, J.D. (2007). *Wind Loading of Structures*, 2ee. New York: Taylor & Francis.

Liu, H. (1991). *Wind Engineering: A Handbook for Structural Engineers*. Englewood Cliffs, NJ: Prentice Hall.

Mehta, K.C. and Coulbourne, W.L. (2013). *Wind Loads*. Reston, VA: American Society of Civil Engineers.

Parkinson, G.V. and Brooks, N.P.H. (1961). On the aeroelastic instability of bluff cylinders. *Journal of Applied Mechanics* 28: 252–258.

Parkinson, G.V. and Smith, J.D. (1964). The square prism as an aeroelastic non-linear oscillator. *Quarterly Journal of Mechanics and Applied Mathematics* 17: 225–239.

Sadraey, M. (2009). *Aircraft Performance Analysis*. VDM Verlag Dr Muller Refer chapter 3.

Schlichting, H. (1968). *Boundary-layer Theory*. New York: McGraw-Hill.

Selvam, R.P. (1990). Computer simulation of wind load on a house. *Journal of Wind Engineering and Industrial Aerodynamics* 36: 029–1036.

Selvam, R.P. (2017a). CFD as a tool for assessing wind loading. *The Bridge and Structural Engineer* 47 (4): 1–8.

Selvam, R.P. (2017b). *Structural Dynamics and Loading*. Linus Learning.

Selvam, R.P. and Holmes, J.D. (1992). Numerical simulation of thunderstorm downdrafts. *Journal of Wind Engineering and Industrial Aerodynamics* 44: 2817–2825.

Selvam, R.P. and Millett, P.C. (2003). Computer modeling of tornado forces on buildings. *Wind & Structures* 6: 209–220.

Selvam, R.P. and Smith, J.W. (1989). Computer simulation of wind flow over steep hills. In: *Recent Advances in Wind Engineering*, vol. I (ed. T.F. Sun), 194–201. Pergmon Press.

Selvam, R.P., Govindaswamy, S., and Bosch, H. (2001). Aeroelastic analysis of bridge girder section using computer modeling. Final report: MBTC FR-1095, Mack-Blackwell National Rural Transportation Study Center, University of Arkansas. http://ntl.bts.gov/lib/11000/11100/11186/1095.pdf (accessed 7 March 2022).

Selvam, R.P., Govindaswamy, S., and Bosch, H. (2002). Aeroelastic analysis of bridges using FEM and moving grids. *Wind & Structures* 5: 257–266.

Simiu, E. and Miyata, T. (2006). *Design of Buildings and Bridges for Wind: A Practical Guide for ASCE-7 Standard Users and Designers of Special Structures*. John Wiley & Sons.

Simiu, E. and Scanlan, R.H. (1986). *Wind Effects on Structures: An Introduction to Wind Engineering*. New York: Wiley.

Strasser, M.N. and Selvam, R.P. (2015). A review of viscous vortex tangential velocity profiles for application in CFD. *Journal of the Arkansas Academy of Sciences* 69: 88–97.

Stull, R.B. (2017). *Practical Meteorology: An Algebra-based Survey of Atmospheric Science, Version 1.02*. University of British-Columbia.

Sumer, B.M. and Fredsøe, J. (2006). *Hydrodynamics Around Cylindrical Structures*. New Jersey: World Scientific Pub.

Warburton, G.B. (1976). *The Dynamical Behavior of Structures*. Oxford: Pergamon Press.

Yousef, M.A.A., Selvam, R.P., and Prakash, J. (2018). A comparison of the forces on dome and prism for straight and tornadic wind using CFD model. *Wind & Structures* 26: 369–382.

5

CFD for Turbulent Flow

We will apply computational fluid dynamics (CFD) for different wind engineering problems in this chapter. The problems are classified as steady and unsteady problems. The steady problems will not take much computer time and the unsteady problems will take extensive computer time. The flow over 2D building and flow over hill does not take much computer time because of 2D modeling as well as interest in steady state nature of the solution. Same way 3D building without inflow turbulence will not take much computer time. Flow over square and circular cylinder takes more computer time. The 3D flow over cube with inflow turbulence will take much larger computer time. From physical point of view, the steady problems give mean pressures reasonably well, and the unsteady problems predict time-varying pressure accurately. For the flow over cylinder, we will use direct simulation for low Re (Reynolds number). For building study, we will use large eddy simulation (LES).

5.1 Mean and Peak Pressure Coefficients from ASCE 7-16 and Need for CFD

The ASCE 7-16 provides mean and peak pressure coefficients C_p for practitioners to use for design of buildings. Here, pressure coefficients $C_p = (P - P_{ref})/(\rho U_{ref}^2/2)$, where P is the pressure at a point on a building, U_{ref} is the reference velocity far away from the building, and P_{ref} is the reference pressure at that point. The pressure coefficients are non-dimensional numbers, and in this way we can compare very easily the level of forces produced for different geometric shapes. As an example, for a square 2D building and cubical 3D building, we will get the C_p from ASCE and compare them with CFD to estimate the capability of CFD as well as to develop understanding of how code works. In this case, 2D and 3D building values are the same for ASCE 7-16. In the ASCE, the main wind force resisting system (MWFRS) C_p are mean pressures and components & cladding (C&C) C_p are peak pressures.

Computational Fluid Dynamics for Wind Engineering, First Edition. R. Panneer Selvam.
© 2022 John Wiley & Sons Ltd. Published 2022 by John Wiley & Sons Ltd.

ASCE Pressure Coefficients C_p from ASCE 27.3.1

MWFRS or mean C_p:

Windward wall $C_p = 0.8$, leeward wall $C_p = -0.5$, sidewalls $C_p = -0.7$, roof $C_p = -1.3$ to -0.7

Components & cladding or peak C_p (ASCE7-16-30.3.1): corner walls: -1.4, roof: $C_p = -3.2$

We will use these values to compare with CFD and wind tunnel (WT) pressure coefficients in the later sections.

Turbulent Flow Effect on Peak Pressures on Building

In the case of WT or CFD, the pressure on the building will be varying in time, and hence the mean and peak pressures have to be calculated from time-dependent records. Generally, the wind flow in the WT will be turbulent. Whereas in CFD, if turbulence is not introduced at the inflow, the turbulent effect on building is not completely modeled. The mean value computed using CFD for a flow without inflow turbulence may be comparable to WT or field measurements. But, the peak pressure will have large error. For accurate computation of peak pressure in CFD, one has to introduce inflow turbulence. There are several challenges in this regard and this is the current research. We will discuss different methods of introducing inflow turbulence in the later part of this chapter.

When turbulence is introduced at the inflow, the peak pressure coefficient C_p may be as high as -7 compared to mean C_p of -2.0 for a roof edge as shown in Figure 5.1a. This high C_p is from field measurement as reported in Richards et al. (2007) and Richards and Hoxey (2012). The corresponding WT peak C_p is -3 from Figure 5.1a. The WT peak C_p of -7 at the corner of the Texas Tech University (TTU) building is far lesser than -18 from field as shown in Figure 5.1b. Currently analytical methods are developed to improve the WT measurement as reported in Mooneghi et al. (2016) and that also shown in Figures 5.1b. The reason for the difference between WT and field measurements is due to the lack of energy in the low-frequency region of the wind spectrum as shown in Figure 5.2.

In the first study, inflow turbulence will not be considered for 2D and 3D flow over building modeling. The main reason for not considering for this study is to reduce computer time and get familiar with CFD modeling. When the inflow turbulence is considered, the computer time will be increased at least by one order of magnitude. For some works, it may take several days of computing and one may need high-performance computing or parallel computing.

5.2 Procedure for CFD Modeling

For CFD modeling of any wind engineering problems, the following procedure needs to be followed:

1) **Mathematical Modeling**: For a given physical problem, the governing equation, boundary condition (BC), and initial conditions have to be determined. Different physical parameters have to be collected and computational domain has to be resolved. If any

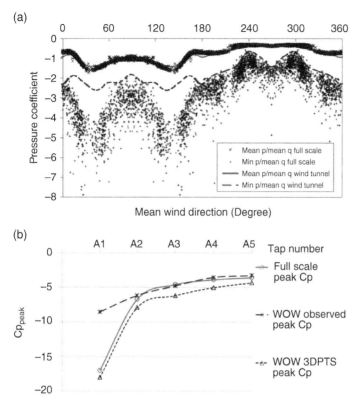

Figure 5.1 Comparison of field and wind tunnel pressure coefficients: (a) Silsoe building field-tap#6. *Source:* Richards et al. (2007)/with permission of Elsevier. (b) Peak pressure coefficient on the roof of TTU building. *Source:* Mooneghi et al. (2016)/with permission of Elsevier.

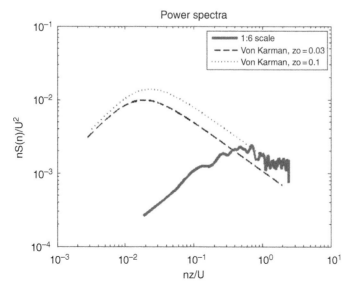

Figure 5.2 Lack of energy in the WT wind spectrum compared to filed measurement for low-rise building. *Source:* Taken from Moravej (2018).

mistake is made in this step, the error goes all the way to the end. For example, for the given problem, one should know the H_{ref}, U_{ref}, Re, P_{ref}, inflow parameters, and turbulence model to be used for incompressible viscous flow. Here, H_{ref} is the reference height of the building or some characteristic length, U_{ref} is the reference velocity, and P_{ref} is the reference pressure. If the flow is assumed to be inviscid, then the corresponding BC has to be determined.

2) **Preprocessing or Grid Generation**: In this step, the flow region has to be discretized with nodes and elements as in finite element method (FEM). The nodes have to be distributed properly to have a better solution. This step may be the most time-consuming one in practical application. For example, to study flow over an airplane, one may spend months to get the grid around it. Several grid generator programs are available for sophisticated modeling. In this text, we consider equal or unequal spacing in a rectangular grid system. This is also called structured grid. This makes it easier to generate the grid. In the FEM classes, detailed discussion of unstructured grid is introduced. For more detail on different grid system, one can refer to Selvam (2020). Strasser et al. (2016) used unstructured grid for flow around circular cylinder, and Selvam et al. (2002) used unstructured grid for flow around a bridge. The generated grid can be visualized using Tecplot, ParaView, etc. The ParaView is an open-source software. Some introduction to Tecplot is provided in Appendix A. Finally, one has to decide what are the data need to be stored or retrieved at each step for processing in the post-processing step.

3) **Running CFD Model**: In this step, the flow solver will solve the Navier–Stokes (NS) equations in time. Here, one has to choose explicit or implicit solver (SIMPLE, Fractional Step, SOLA, etc.), method to approximate time term, convection, and diffusion terms, etc. Method to solve the momentum equation and pressure equation has to be decided. This step is the most computer time-consuming step.

4) **Post-Processing**: In this step, the developed data will be analyzed with many ways. For 2D and 3D problems, one can use contour and vector plot for visualization. Time-dependent data can be used for calculating wind spectrum plot, animation of the flow, etc. Making movies using the data created is an art. One has to decide properly what you see and how to show it to others. Procedure to make animation using Tecplot is given in Appendix A. If proper data management is not done, one can produce extensive data. For inflow turbulence effect on building, one may run several thousands of time step.

5.3 Need for Nondimensional Flow Modeling

In all our work, instead of using actual dimensions in the CFD modeling, nondimensional flow region and BCs will be used. This is an important step in CFD modeling. The advantages of using nondimensional study instead of dimensional modeling:

1) The NS equation is a nonlinear equation and the equations are solved by iterative procedures. In the iterative procedure, the error or residue in each equation is reduced to 0 by iteration. The Guss–Seidel (GS) and Successive-Over-Relaxation (SOR) iterations are explained in Chapter 3. When dimensional numbers are used, the error or RNORM may

be a very big number. If the residue norm or RNORM goes beyond $10^{\pm 25}$, the computer cannot handle these numbers. By scaling the dimensions, the number will be small. The scaling is done by dividing the distances by H_{ref} and velocity by U_{ref}. The detail of non-dimensionalizing the NS equation is discussed in Chapter 2. Reverse approach will be used to compute dimensional values from nondimensional numbers.

2) For the incompressible flow, the only nondimensional number in the NS equation is Reynolds number Re ($U_{ref}H_{ref}/\nu$). By varying the Re, one can understand the flow behavior very well.

3) By solving nondimensional NS equation, the pressure coefficient is computed as twice the nondimensional pressure P^*. That is in the computation $U_{ref} = 1$ and $(P - P_{ref})/\rho = P^*$ and hence $C_p = (P - P_{ref})/(\rho U_{ref}^2/2) = 2P^*$.

5.4 Flow Over 2D Building and Flow Over an Escarpment

5.4.1 Program uvps3.f, to Study Flow Over a Hill or Flow Around a Building

Modeling Details and Boundary Conditions: The program solves the 2D Navier–Stokes equations on a nonstaggered grid. A constant grid spacing of h is considered in both x and y directions. The inlet value is kept a **constant U velocity in the x direction and zero V velocity in the y direction**. In the actual boundary layer, it will be a logarithmic profile as discussed in Chapter 4. Using **the inlet velocity as the reference velocity**, the nondimensional value of U at the inlet will be 1 (U/U_{ref}). Similarly, we will take the **building height (H_{ref}) to be the reference height** and hence the nondimensional value of the building height will be 1 (x/H or y/H). The outlet is considered to be free outflow, which means the normal derivative of velocity is 0. On the wall, no-slip condition is used (U = V = 0) and at the top boundary U is kept 1 and V as 0. The details of the BCs are shown in Figure 5.3. At each time step, the pressure is initialized to 0 at the outlet point JM-5. Here, IM and JM are the number of grid points in the x and y directions. The maximum points that can be used in the x and y directions are **151 × 101,** respectively. The time step is controlled by Courant–Friedrichs–Lewy condition (CFL) criteria of dt < h/U_{max} for convection term, and for diffusion term the stability criteria is dt < $h^2/(4\nu)$. Here, U_{max} is the maximum velocity in the flow region. The program considers constant time step of dt. The initial condition for the flow is U = 1 and V = 0 for all the interior points.

Further Improvements: The inlet velocity can be improved for logarithmic profile as it is in an atmospheric boundary layer and turbulence can be modeled by LES instead of increased viscosity. The grid spacing can be changed into variable grid spacing.

Methods to Include Obstacle: The building or escarpment is considered by the penalty method. This means that the momentum equations are specified with zero velocity at the building location and the continuity equation is solved for the whole domain including building. This is not an accurate method and the pressure on the building will have much error. The obstacle is identified by the one-dimensional array (KH) along the x-axis. The KH array is filled with 1 for flat wall and more than 1 for obstacle. As an example at some distance say I = 9, if the obstacle height is say fourth or fifth point from bottom as shown in

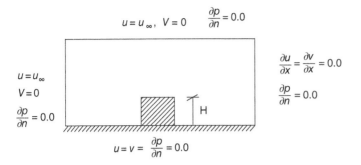

Figure 5.3 Boundary conditions for 2D flow over obstacle.

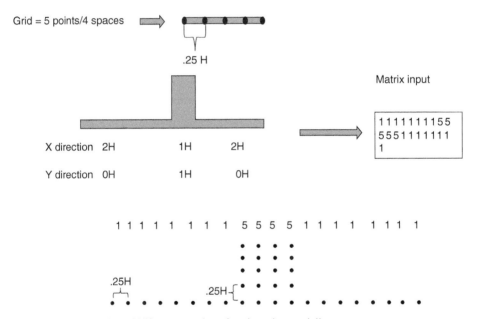

Figure 5.4 Illustration of KH array numbers for obstacles modeling.

Figure 5.4, then KH(9) = 5. The KH array has to be provided for all the points in the x direction as shown in Figure 5.4.

The NS equations are solved by fractional step procedure by Chorin (1968). The momentum equations are solved explicitly at the interior points and the pressure equation is solved by SOR. Only pressure at the interior points is solved and the outer points are interpolated from the interior points. The convection part is approximated by the upwind procedure. This is a very diffusive method and hence unsteady nature of the flow will be damped. This part can be improved in the future. Overall, the program is suitable for steady flow calculation. Since the CFD code **uvps3** does not consider the turbulence effect, the viscosity value can be increased as in Hirt et al. (1978). Since upwind is very diffusive, this should give enough numerical stability.

Mixing Length Theory for Boundary Layer Flow or Modification to Kinematic Viscosity: For turbulent boundary layer, the mixing length theory as discussed in Schlichting (1968) or Versteeg and Malalasekera (2007) can be used. The k–ε turbulence model also gives similar results as reported in Selvam (1990). The turbulent eddy viscosity can be calculated as:

$$\nu_t = 0.4(y + y_0)u^*$$

where y is the distance from wall, y_0 is the roughness coefficient, and u^* is the frictional velocity. This can be added to the viscosity. In this code, only constant viscosity is used.

Limitations:

1) Not considering any turbulence model.
2) Equal spacing is considered. Unequal spacing may help to use the grid optimally.
3) May not resolve the boundary layer with proper grid refinements. Hence, local flow variation or local recirculation may not be captured.
4) The turbulence effect or flow effect may be 3D in many situations and hence 2D flow may not capture the relevant issues.
5) Not considering inflow turbulence.

This computer model is better than potential or inviscid flow model. The difference between inviscid flow and viscous flow is that the velocity at the wall is zero in the case of viscous flow. This can predict the flow recirculation effect.

The program needs input data, and it is given in the input file name called **uvp3-i.txt**. The final output is written in **uvp3-p.plt**. This has u, v, and p values for the final step.

Example 5.1   Flow over a Square Building

To perform this analysis, the nondimensional NS equation is considered. The reference values considered are U_{inlet} and building height H. Here, viscosity input becomes 1/Re (VIS = 1/Re) where Re is the Reynolds number. The Reynolds number is defined as $Re = U_{ref}H_{ref}/\nu$.

The flow region considered is shown in Figure 5.5. The computational region is 15H × 5H (XL*YL). The building is considered at 5H from the inlet. We keep $H_{ref} = 1$. If we say we are discretizing the H_{ref} by four spacing, then h = $H_{ref}/4 = 0.25H_{ref} = 0.25$. Then number of points in the x direction will be IM = $15H_{ref}/(H_{ref}/4) + 1 = 61$. Since equal spacing is used in both directions in this program, the number of points in the y direction will be JM = 5 × 4 + 1 = 21. Hence, five points or four spacing is used for the building. At this time, Re = 1/0.01 = 100 is used. The relaxation factor (RF) value should be less than 2. Optimum value can give you very less iteration. The maximum SOR iteration is kept as **niter** = 500 inside the program. If the iteration is 500 or more, one has to reduce the time step **DT** to get proper convergence or increase the **NITER** value. The **NITERT** should be long enough to achieve convergence for steady state calculations. Assuming the U_{max} may be around 2 units, one can calculate the dt = $h/U_{max} = 0.25/2 = 0.125$ units for convection stability criteria. For diffusion, dt < $h^2/4\nu = 1/(4 \times 16 \times 0.01) = 1.6$ units. In the following provided input data, dt = 0.001, which is far less than CFL criteria of 1, and hence it should be fine for convergence.

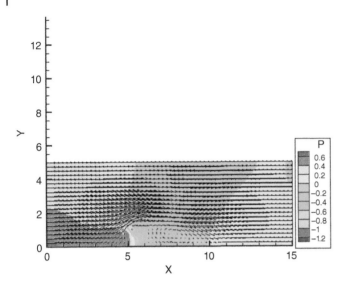

Figure 5.5   Flow region with building.

Input Details in **uvp3-i.txt** for flow around the building:

```
READ(1,*)IM,JM,XL,VIS, UIN, DT,NITERT,NITER,RF
READ(1,*)(KH(I),I=1,IM)
IM    #OF POINTS IN X (Max. Allowed IM=151)
JM    #OF POINTS IN Y (Max. allowed JM=101)
XL    LENGTH OF THE DOMAIN IN X
VIS   VISCOSITY (1/Re)
UIN   U VELOCITY AT INLET
DT    TIME STEP CONTROLED BY CFL NUMBER
NITERT    # OF TIME STEP TO BE CONSIDERED
NITER     MAX. #OF ITERATION ALLOTED FOR SOR
RF    RELAXATION PARAMETER <2
**********
KH    HEIGHT FROM GROUND 1 OR MORE FOR IM POINTS
```

Sample data for flow over square building:

```
61,21,15.,.01,1.0, 0.001,5900,500,1.5
1   1   1   1   1   1   1   1   1   1
1   1   1   1   1   1   1   1   1   1
5   5   5   5   5   1   1   1   1   1
1   1   1   1   1   1   1   1   1   1
1   1   1   1   1   1   1   1   1   1
1   1   1   1   1   1   1   1   1   1
1
```

Output Details: Screen output and **uvp3.plt file.**

The Screen Output at Each Time Step: Number of time step, # of iteration for SOR, error norm, and velocity maximum in the domain.

The error norm is calculated as absolute sum of the error at each node. This error is checked at every iteration. If this error is larger than the tolerance value ($IM^*JM^*10^{-5}$), then further iteration continues until convergence or allowed maximum number of iterations (**NITER**).

The uvp3.plt contains u, v, and p for visualization.

Interpretation of Computed Results: The pressure at the inlet has a positive value of 0.4–0.6 units as shown in Figure 5.6. This may be due to inlet close to the building. The other reason may be due to 2D modeling also. In 3D modeling, the flow can go around the building in addition to going over the building. Hence, for 3D modeling, 5H distance may be fine. **Since the pressure far away from the building should be zero, the particular modeling is not proper and this could be improved.** Usually the pressure from the computer model is converted to pressure coefficient C_p and this is a nondimensional number as follows:

$$C_p = (P - P_{ref})/(\rho U^2/2)$$

Here, P_{ref} is the reference pressure. In our case, this can be taken as the P at the outlet. This P is 0 or close to 0 in our case. Then, $C_p = 2P$ on the building.

To understand the flow, only vector diagram is plotted in Figure 5.6. Here, the building is clearly seen as the blank region. The P_{min} and P_{max} reported in the CFD model are −1.29 and 0.74, respectively. The variation of P is shown in Figure 5.7b. Hence, the roof C_p from CFD are $C_{pmin} = -2.58$ and $C_{pmax} = 1.48$. From ASCE 7-16, we observed MWRFS C_p on roof = −1.3 to −0.7 and C&C $C_p = -3.2$. Further modeling with more extended region on the upwind side and running for much longer time may help to understand the performance of 2D modeling.

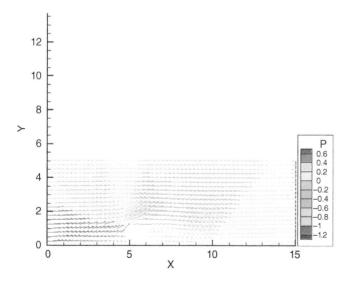

Figure 5.6 Velocity contour with pressure values as the color for velocities.

(a)

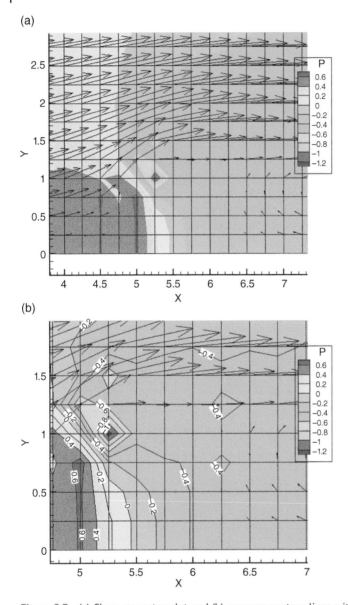

(b)

Figure 5.7 (a) Close-up vector plot and (b) pressure contour lines with values.

Example 5.2   Flow Over an Escarpment

The flow region considered is shown in Figure 5.8 for flow over an escarpment. The slope of the hill is considered to be $45°$. The height of the hill is $H = 1$. The computational region is $15H \times 5H$. The grid spacing $h = XL/(IM − 1) = 15/60 = 0.25$. The escarpment starts from 7.25H or 29th point from the inlet. In five points, the escarpment reaches the height of 1 unit. At this time, $Re = 1/0.01 = 100$ is used.

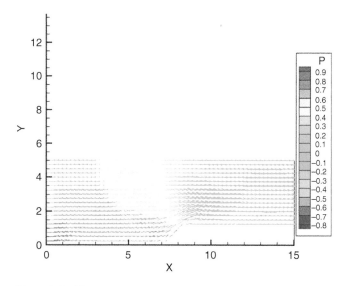

Figure 5.8 Flow over an escarpment.

Sample data flow over escarpment.

```
61,21,15.,.01,1.0,0.001,5900,500,1.5
1   1   1   1   1   1   1   1   1   1
1   1   1   1   1   1   1   1   1   1
1   1   1   1   1   1   1   1   1   1
2   3   4   5   5   5   5   5   5   5
5   5   5   5   5   5   5   5   5   5
5   5   5   5   5   5   5   5   5   5
5
```

For further understanding of U and V velocity variation in the computational domain, corresponding contour plots are shown in Figures 5.9 and 5.10.

The ASCE 7-16 gives a table to calculate the velocity on an escarpment. The velocity over the escarpment can be calculated using the formula: $1 + K_1K_2K_3$

where

K_1 account for shape. Here, we will choose escarpment.
K_2 x-position from the crest
K_3 y-position

The ASCE 7-16 enhances the pressure for hill effect by multiplying by the factor $K_{zt} = (1 + vK_1K_2K_3)^2$ because P is proportional to U^2. Since we compare velocity only, we compare with $1 + K_1K_2K_3$.

Here, L_h = width at half-height = 0.5 and H = 1.0.

$H/L_h = 1/0.5 = 2$. Take $H/L_h > 0.5$ as 0.5 and $K_1 = 0.43$.

Substitute $L_h = 2H$ for K_2 and K_3 computation.

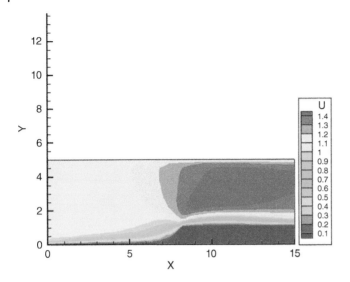

Figure 5.9   U velocity contour plot.

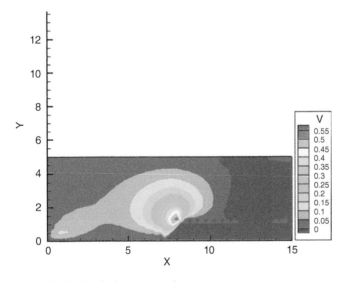

Figure 5.10   V velocity contour plot.

$K_2 = 1.0$ for $x/L_h = 0$.

For y/L_h, K_3 is given. Hence, $K_1 K_2 K_3 = 0.43 K_3$.

Using the earlier mentioned information, the increase in velocity above the crest of the escarpment is calculated and reported in Table 5.1.

The CFD velocities (U and V) are retrieved from the CFD output and are reported in Table 5.2. Using this value, the magnitude of the velocity is calculated and reported as V_{mag}.

Table 5.1   Increase in velocity due to escarpment from ASCE 7-16, Table 26.8-1.

$y/L_h = y/2$	0	0.1	0.2	0.3	0.4	0.5	0.6	0.7	0.8	0.9	1.0
y	0	0.2	0.4	0.6	0.8	1.0	1.2	1.4	1.6	1.8	2.0
K_3		1.0	0.78	0.61	0.47	0.37	0.29	0.22	0.17	0.14	0.11	0.08
$K_1K_2K_3$		0.43	0.34	0.26	0.2	0.16	0.12	0.09	0.07	0.06	0.05	0.03

Source: ASCE 7-16/ASCE.

Table 5.2   Magnitude of the velocity above the escarpment.

$y - 1$	0.0	0.25	0.5	0.75	1.0	1.25	1.5	1.75	2.0
U	0.0	0.78	1.23	1.32	1.37	1.37	1.36	1.35	1.34
V	0.0	0.43	0.36	0.38	0.31	0.28	0.22	0.21	0.16
V_{mag}	**0.0**	**0.89**	**1.28**	**1.37**	**1.40**	**1.40**	**1.38**	**1.37**	**1.35**
ASCE	**1.43**	**1.32**	**1.23**	**1.17**	**1.12**	**1.09**	**1.07**	**1.06**	**1.03**

Source: ASCE 7/ASCE.

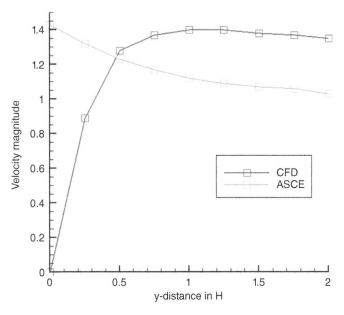

Figure 5.11   Velocity with height for CFD and ASCE.

The velocities from the ASCE 7 are also reported in the same Table 5.2 using the information in Table 5.1. From the table as well as from Figure 5.11, one can see that the ASCE velocities are higher close to the ground compared to CFD. This is because ASCE values may be based on inviscid calculation. As the height increases, the velocity increases up to 1.25H and then decreases for CFD as shown in Figure 5.11. Further improved values can be obtained using refined grids.

The ASCE values are applicable for milder slopes because the potential theory is valid when there is no flow separation. Since the slope is 45°, the ASCE values are not that reliable. A correction may be necessary for the ASCE at higher slopes.

5.5 Pressure on the Texas Tech University (TTU) Building Without Inflow Turbulence

Field measurements and several WT studies were conducted on a cube building or Silsoe building and TTU building. Hence, these two become a good benchmark problem to illustrate CFD applications. A cubic building is attractive because the grid resolution is much less compared to TTU building. One drawback with the cubic or Silsoe building data is that there is no pressure on the side as per Mooneghi et al. (2016). The TTU field measurements reported in Levitan et al. (1991) and Moravej (2018) are also available for CFD study. Many WT studies are conducted for the TTU building and they are available in the open literature. So, we will consider TTU building for our study.

The earliest work on flow over a building is reported in Hirt and Cook (1972) and Caretto et al. (1972). Hirt and Cook (1972) showed the fractional-type solution technique for flow around buildings. Mainly, this work focused on pollutant transport around building. They used SOLA solver for pressure correction. They used $15 \times 15 \times 15$ mesh for their work. Caretto et al. (1972) used $10 \times 10 \times 10$ unequal spacing grid for the entire computational region. This is the same paper where SIMPLE and SIVA procedures were introduced. The same group improved the work by using k–ε turbulence model, and the details are reported in Vasilic-Melling (1976) PhD thesis. In this work, 2D and 3D building pressure were compared with WT measurements. Selvam (1992) compared TTU field measurements with steady CFD computations. The mean pressures were well compared at several locations. Later Selvam (1997) used LES with some form of inflow to compare the field mean and peak pressures over TTU building. Without inflow turbulence, with refined grid close to the building reported in Selvam (2010) showed flow separation on the roof and produced peak pressures equal to 1 : 50 scale WT measurements. There are several other researchers who contributed in this field, and some are reviewed in Selvam (2010, 2017). Any computation with inflow turbulence takes extensive computer time, and this will be discussed in later section in this chapter. Here, we will investigate the pressure on building without inflow turbulence. This means only mean velocities are given as input value at the inlet.

5.5.1 Mathematical and Numerical Modeling

LES is considered as the turbulence model. The governing equations and the solution procedure for the NS equations are available from Selvam (1997). The NS equation is solved by fractional step method and the variables are stored at the node points. Hence, nonstaggered grid system is used. The Crank–Nicolson method is used to approximate the unsteady and convection term. The space variables are approximated by the central difference method.

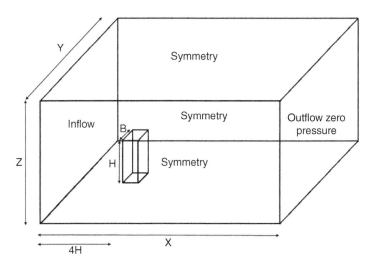

Figure 5.12 Computational region with boundary conditions.

The momentum equations are solved by line iteration (Tridiagonal matrix algorithm [TDMA]), and the pressure equation is solved by preconditioned conjugate gradient (PCG) method. The detail of numerical implementation is reported in Selvam (1996). To satisfy the NS equations completely at each time step, a maximum subiteration of 10 is kept for convergence.

The computational region is shown in Figure 5.12. On the building and on the ground, no-slip condition is used. That is, the velocities are zero and the normal derivative of the pressure is zero. The logarithmic wind profile at the inlet cannot be simulated in the computational domain using H/16 grid because of lack of grid resolution. To rectify that problem, the law of the wall BC discussed in Chapters 2 and 3 will be introduced. On the side and top boundaries, symmetric BC is used ($U_n = 0.0$, other velocities $dU_i/dn = 0.0$ where n is the normal direction). At the inlet, mean velocity is a logarithmic profile for U in the x direction and other velocities V and W are zero.

At the outlet, convective BC is used. The benefit of the convective BC over Neumann BC is discussed in Sohankar et al. (1998). That is for any velocity U_i in the x direction:

$$dU_i/dt + U_n dU_i/dn = 0.0$$

The initial condition is considered to be mean velocity at the inlet. At the outlet, the pressure is specified to be zero on the top one-fourth region and bottom region normal derivative of p is zero.

5.5.2 Detail of the TTU Building and the Computational Region

The dimension of the TTU building is $2.25H \times 3.375H \times H$, where H is 4 m. The reference height for computation is H. The flow is considered to be along with the shorter length (2.25H) of the TTU building. The reference velocity U_H at the building height is considered to be 8.6 m/s and the roughness length of the ground z_0 is 0.024 m as reported in

Selvam (1992). From this, we can calculate the Reynolds number Re $= UH/\nu =$ $8.6 \times 4/1.52 \times 10^{-5} = 2.3 \times 10^{6}$. In the input of the program, we use **VISC = 1/Re =** $\mathbf{4.0 \times 10^{-7}}$ by rounding the number. For TTU-type bluff body, the flow is Re independent because the flow separation happens at the edge of the windward roof. The corresponding nondimensional value is $z_0 = 0.024/4 = 0.006$ and this is given as $\mathbf{C_2}$ **in the input preparation**. Since the reference velocity is at the building height, for nondimensional calculation this value is 1. For logarithmic inflow profile, the coefficient C_1 is calculated as follows:

$$C_1 \ln\left((z + C_2)/C_2\right) = 1.0, \text{then } \mathbf{C1 = 1/\ln(1.006/0.006) = 0.195}$$

The computational region around the building has to be such that the outer BCs should not introduce much error on the pressure over the building. In the field, the free wind all around the building is there. This means the outer boundary can be extended to any length, but they are computationally intensive. So, from numerical experiment of the previous works like Selvam (1992, 1997), we can say that 4H and 7H on the windward and leeward part of the building along the x direction, 3H on either side of the building in the y direction, and 4H above the building in the z direction can be used as a start. This amounts to $13.25H \times 9.375H \times 5H$ as the computational region as shown in Figure 5.13.

5.5.3 Grid Generation

The solution region is $13.25H \times 9.375H \times 5H$. The solution region has to be discretized with grid points as we discussed in the Chapter 3. For practical problems, discretization of the flow region may be very complex and grid generation may be the most time-consuming step. For aeronautical problems, there are people who work full time for a month or more to design the grid. In this work, simple orthogonal grid will be used for illustration. This eliminates the challenges in grid generation. The whole region is discretized with an equal

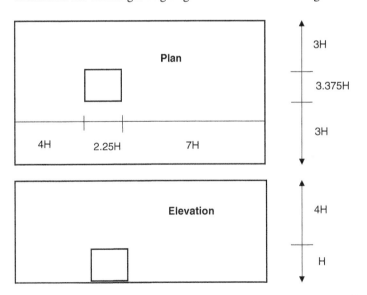

Figure 5.13 Plan and elevation of the computational region shown in H.

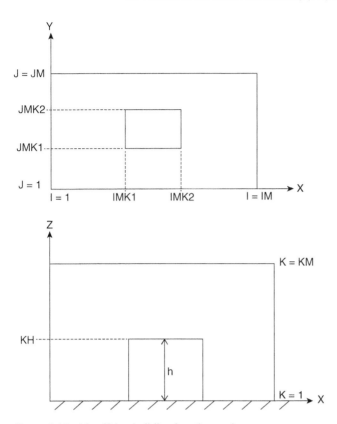

Figure 5.14 Identifying building location nodes.

grid spacing of H/16. This means there are 16 grid spacing in H. This makes it easier to give input for the program. The number of grid points in the x, y, and z directions are called IM, JM, and KM, respectively. They are calculated as:

$$IM \times JM \times KM = (13.25 \times 16 + 1)(9.375 \times 16 + 1)(5 \times 16 + 1)$$
$$= 213 \times 151 \times 81 = 2\ 605\ 203 \text{ or } 2.6 \text{ million grid points.}$$

The beginning and end nodes of the building location in the x (IMK1, IMK2), y (JMK1, JMK2), and z (KH) directions are provided as input and are shown in Figure 5.14. Similarly, the location of the building nodes in the x, y, and z directions are calculated as:

$$IMK1, IMK2 = (4 \times 16 + 1 = 65), (6.25 \times 16 + 1 = 101)$$
$$JMK1, JMK2 = (3 \times 16 + 1 = 49), (6.375 \times 16 + 1 = 103)$$
$$KH = 16 + 1 = 17.$$

5.5.4 Time Step and Total Time to Run

The computer time step is calculated based on the grid spacing. A nondimensional time step of 0.02 is used. This will make the CFL condition less than 1. This is arrived by assuming the maximum velocity around the building may be 2 units and $dt < h/U_{max} = 1/(16 \times 2) = 0.03125$ units. The total time to run is determined by the sum of initial time for the flow to settle and the unsteady nature of the flow developed. We will plot the pressure at different times and see the variation for this. As a start, let us keep 20 time units.

Computer time: 194 minutes or about 3 hours in a personal computer.

5.5.5 Details of Program yif2.f

This program can perform 3D-CFD calculations for flow over building using inflow or no-inflow turbulence. The details of computer modeling, and initial condition and BC are discussed in the previous section. In this part, we will illustrate how to prepare input, and to interpret and visualize the output data for the case of no-inflow turbulence. In the later section, we will use inflow turbulence based on Yu et al. (2018). Equal grid spacing is used in this program. The maximum number of outer iterations is kept as 10.

5.5.6 Files Needed to Run the Program

To run the program, one needs **yif2.exe**, **yif-i.txt**, and **char.txt**. The **char.txt** has 999 names of the movie files stored. In case someone wants to make movies, they can activate this part and they should make sure the movie files are not more than 200.

5.5.7 Input Data File: yif-i.txt

The input data is illustrated with the data discussed in the previous Sections 5.5.2–5.5.6.
Input reading detail:

```
READ(5,*)IM,JM,KM,IMK1,IMK2,JMK1,JMK2,KH,DTT,TTIME
READ(5,*)C11,C2,IPLOT,VISC,XL3,YL3,ZL,DF,IFS,IFE,HREF,UREF,INFLT
```

Line 1: `READ(5,*)IM,JM,KM,IMK1,IMK2,JMK1,JMK2,KH,DTT,TTIME`

IM	# of points in x, IM = 213
JM	# of points in y, JM = 151
KM	# of points in z, KM = 81
IMK1	Point from where building starts in x (refer Figure 5.13 for details), IMK1=65
IMK2	Point from where building ends in x, IMK2 = 101
JMK1	Point from where building starts in y, JMK1 = 49
JMK2	Point from where building ends in y, JMK2 = 103
KH	Point up to which building is there, KH = 17
DTT	time step dt used, DTT = 0.02
TTIME	Total time to run the program, 20.0

Line 2:
`READ(5,*)C11,C2,IPLOT,VISC,XL3,YL3,ZL,DF,IFS,IFE,HREF,UREF,INFLT`

C11 coefficient for the log law profile $\{C_{11} = u(z = 1)/\ln[(1 + C_2)/C_2]\}$, C11 = 0.195 from 5.5.2

C2 nondimensional roughness length ($C2 = z_0/h$, where h is the reference length = 0.006)

IPLOT   frequency of writing movie files-5000
VISC   equal to $1/Re = 4.\times10^{-7}$
XL3   domain length in $x = 13.25$
YL3   domain length in $y = 9.375$
ZL   domain length in $z = 5.0$
DF   frequency width dn in $Hz = 0.1$
IFS & IFE   starting and ending range of n#. Here, the frequencies
                        n in Hz: 2.86, 4.0
HREF, UREF   reference height in m and velocity in m/s: 4.0, 8.6
INFLT   If INFLT = 1 consider inflow turbulence. INFLT = 0, no-inflow
                        turbulence: 0

For DF = 0.1, IFS = 2, IFE = 20

$n1 = DF(2^*IFS - 1)/2 = 0.1 \times 3/2 = 0.15\,Hz$ $f_1 = n_1 H/U = 0.15(4/8.6) = 0.07$
$n2 = DF(2.^*IFE - 1)/2 = 0.1 \times 39/2 = 1.95\,Hz$ $f_2 = n_2 H/U = 1.95(4/8.6) = 0.91$

From f_1 and f_2, calculate n_1 and n_2. Then, we should determine DF, IFS, and IFE. The details of this calculation are provided in the section on inflow turbulence.

H/16 sample input data
213,151,81,65,101,49,103,17,0.02,20.0
0.195,0.006,5000,4.E-7,13.25,9.375,5.,0.1,2,86,4.0,8.6, 0

From the input data, one can see that the h = H/16 is not provided directly. The h in the program is calculated as h = XL3/(IM − 1). So one should make sure that the IM, JM & KM and XL3, YL3 and ZL are properly related when h is used. If not, the computer model will be different from what you expected.

Program to run in PC or Linux system
If the program is run in a PC, the following command is sufficient for the error at each step to write on screen in a dos or command prompt environment.

>yif2.exe

In the earlier mentioned case, the screen output will be lost if it is writing for a long time. If that needs to be stored in a separate file, the following command can be used:

>yif2.exe>dum.txt

The dum.txt stores the screen output.

In the case of Linux system, the following command is used for long-time running without being logged in:

$nohup yif2.out>dum.txt &

Once the job is submitted, one can log out and check the progress later.

5.5.8 Output Detail

In addition to writing the RNORM values of velocities and pressure at each outer iteration, the following files are written: **yif-o.plt, yif-o2.plt, yif-o3.plt, yif-p.plt, and prcon.plt**. The details of each output will be discussed in each of the following subsections. Here, a brief note is provided to know the purpose for each data file:

1) **yif-o.plt**: time versus pressure at four points along the central line of the building (windward wall at mid-height, windward edge of the roof, central point of the roof, and leeward wall at mid-height) are recorded.
2) **yif-o2.plt**: statistics of pressure coefficients C_p (mean, peak, and rms) along the center line of the building.
3) **yif-o3.plt**: time versus velocity at the building location (U, V, and W), only U velocity at the inlet and pressure at the building location.
4) **yif-p.plt**: final velocity and pressure for the whole domain.
5) **precon.plt**: statistics of pressure coefficients C_p on the wall and roof of the building.

First, one needs to look at the screen writing or the file that has it to make sure there is convergence at each time step. Before plotting peak pressure coefficients and other important ones needed for design or other purposes, one needs to visualize the 3D plot in different forms to make sure the flow is reasonable and the BCs are implemented properly, using **yif-p.plt** file or other time-recorded files.

5.5.9 Screen Writing

```
WRITE(*,40) ISUB,(RNORM(I),I=1,4),time
ISUB        # of momentum and pressure solution iteration
RNORM       absolute sum of the error for u, v, w, p
Time        time at the particular time step
```

[rps@cmfn cwe-book]$ tail dum

```
4    0.716E-02    0.515E-02    0.522E-02    0.191E+01    0.198E+02
4    0.738E-02    0.531E-02    0.542E-02    0.196E+01    0.199E+02
4    0.725E-02    0.517E-02    0.536E-02    0.192E+01    0.199E+02
4    0.740E-02    0.528E-02    0.552E-02    0.197E+01    0.199E+02
4    0.727E-02    0.513E-02    0.544E-02    0.194E+01    0.199E+02
4    0.743E-02    0.524E-02    0.559E-02    0.201E+01    0.199E+02
4    0.728E-02    0.511E-02    0.547E-02    0.200E+01    0.200E+02
4    0.743E-02    0.524E-02    0.559E-02    0.207E+01    0.200E+02
4    0.738E-02    0.522E-02    0.553E-02    0.209E+01    0.200E+02
4    0.729E-02    0.518E-02    0.545E-02    0.209E+01    0.200E+02
```

The earlier given screen shot is at the end of the computer run. The pressure error is higher than the velocity errors. The pressure solution is the one that takes lots of computer time. The convergence criteria for inner and outer iterations are 1×10^{-5}.

5.5.10 File Detail: yif-o.plt

```
WRITE(2,50)time,V(IMK1,J1,KH/2,4),v(imk1+1,j1,kh,4),V(I1,J1,KH,4)
    &,V(IMK2,J1,KH/2,4)
```
Time pressure at each time: windward middle, roof, roof middle, leeward middle

The variation of pressure on the building in time is recorded in time in **yif-o.plt**. The pressure is monitored only at four points along the centerline of the flow direction as shown in Figure 5.15. The initial pressures $(P - P_{ref})/(\rho U_{ref}^2)$ are high and then come to some level of steady state after five time units in this case as shown in Figure 5.16. The pressure on the roof edge varies in time more than other points. The windward wall pressure is slowly decreasing and hence not clear if the mean pressure calculated is a valid one. This file will be more useful when inflow turbulence is used. Overall, we can see that for the flow to settle down with respect to numerical issues takes about 10 time units. This information will be used in calculating the building statistics.

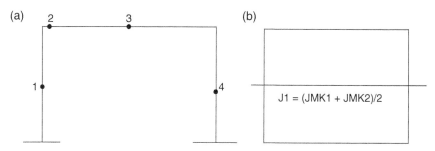

Figure 5.15 Illustration of points where pressure is recorded in time on the building: (a) elevation view and (b) plan view.

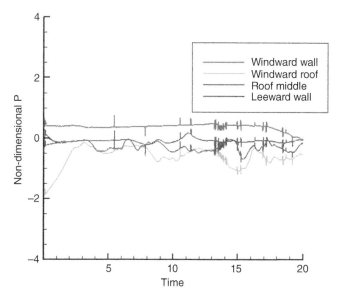

Figure 5.16 Plot from **yif-o.plt** file or time variation of pressure.

5.5.11 File Detail: yif-o2.plt

```
WRITE(4,50)XL2(I),PP2(I,1),PP2(I,2),PP2(I,3),PP2(I,4)
```
XL2   x-distance along the centerline of the building with origin on the roof edge
PP2(I,1)   average pressure coefficient C_p-ave
PP2(I,2)   maximum pressure coefficient Cp-max
PP2(I,3)   minimum pressure coefficients C_p-min
PP2(I,4)   RMS: C_p-Prms

The pressure statistics are calculated from 10s to TTIME for all the points along the centerline of the building as shown in Figure 5.17. The mean and peak pressure coefficients are reported in **yif-o2.plt** file. Here, the reported pressures are pressure coefficient $C_p[2(P - P_{ref})/(\rho U_{ref}^2)]$. As mentioned before, the pressure statistics are calculated from 10 time units to TTIME. For this plot, the statistics may not be completely correct as reported from **yif-o.plt**. The C_p mean, max, and min are reported in Figure 5.18. This plot can be improved by running for a long time and see the time variation performance. From the plot, we can see that even for mean wind flow as inflow, there is time variation of pressure. This happens because of the turbulence generated around the building, and this is

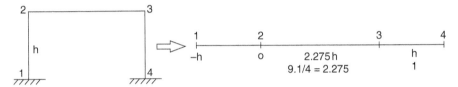

Figure 5.17   Recorded C_p statistics along the centerline of the building.

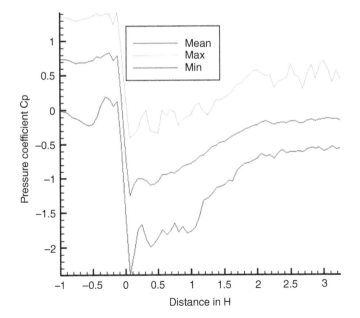

Figure 5.18   Mean, max, and min pressure coefficients along the centerline.

illustrated using **yif-p.plt** file. Selvam (2010) showed that the computed peak pressures are in comparison with 1 : 50 scale WT data for the refined grid they used with mean wind flow at the inlet.

The CFD data file is copied and inserted in the **yif-02-pressure.plt** to compare with WT50 and WT6 data. Here, WT6 means 1 : 6 scale WT measurements and WT50 means 1 : 50 scale WT measurement. The WT6 and WT50 pressure coefficients C_p came from Moravej (2018). The CFD mean and minimum C_p are compared with WT50 and WT6 data in Figure 5.19. The CFD mean C_p values are closer to WT50 than WT6. The CFD peak values

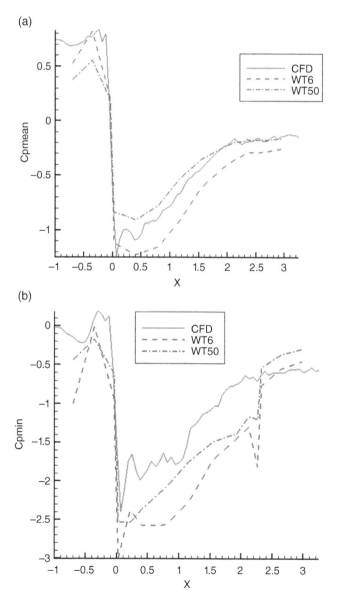

Figure 5.19 CFD pressure coefficient comparison with WT50 and WT6 measurements: (a) C_p mean plot and (b) C_p min plot.

are far less than WT50 measurements. This is expected because there is no inflow turbulence. In the latter section, we will see how CFD compares with WT measurements when the inflow turbulence is considered.

5.5.12 File Detail: yif-o3.plt

```
WRITE(9,50)TIME,V(IMK1,J1,KB,1),V(IMK1,J1,KB,2)
    &,V(IMK1,J1,KB,3),V(1,J1,KB,1),V(IMK1,J1,KB,4)
```

Time	time
(V(IMK1,J1,KB,I1), I1 = 1,3)	3 velocity at the front of the building height when there is no building and if the building is there then they should be zero
V(1,J1,KB,1)	u velocity at the inlet for $z = h = 1$
V(IMK1,J1,KB,4)	pressure at the building location

This file is used to perform spectral analysis of the wind velocity at the inflow and at the building location. If the run is made with building, then velocity at the building location is zero and at the inflow location it is the inflow velocity U(t). In this case, the run is made with building and without inflow turbulence, and hence it is a constant value of U_{ave}. This file data will be useful for inflow turbulence study without building. This part we will consider in the later section. For spectral analysis, one can use the program spd1.f listed in Chapter 2.

5.5.13 File Detail: yif-p.plt

The file **yif-p.plt** is written at the end of the run. This has velocities and pressure at all the points. Hence, this can become a huge file depending on the size of the problem run. One should use the **preplot** option to convert to binary file, and this will reduce the storage and reduces the time to load and unload the data into Tecplot. The detail of converting to binary file is given:

Use **preplot** (**preplot yif-p.plt yip-pe.plt**) option to generate binary file to reduce storage as well as reducing time to load the plot file.

The detail of writing the 3D data for Tecplot visualization is shown:

```
C.....WRITE TECPLOT
      write(7,*)'VARIABLES = "X","Y","Z","U","V","W","P"'
      write(7,*)'ZONE I=',IM, ',J=',JM,',K=',KM, ',F=POINT'
      do k=1,km
      do j=1,jm
      do i=1,im
      write(7,*)x(i),y(j),z(k),(v(i,j,k,i1),i1=1,4)
      end do
      end do
      end do
```

Visualizing in 3D can be done in several ways. Complete 3D visualization is shown as pressure isosurface plot for $p = -0.15$ in Figure 5.20. In the same way, the velocity

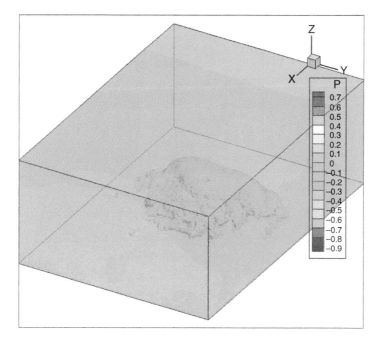

Figure 5.20   3D pressure variation around the building.

magnitude is plotted as isosurface plot for $V_{mag} = 1.35$ and 0.31 in Figure 5.21. The velocity magnitude as well as other variables can be calculated within Tecplot environment using two methods as discussed in Appendix A. From the Figure 5.21, one can see that the velocity magnitude of 1.35 occurs only at certain locations around the building. Due to the building effect, the 1.35 isosurface on the top dips below above the building as shown in Figure 5.21a. The isosurface plot for 0.31 in Figure 5.21b shows the variation around the building as well as far away from the building. Visualization is some level art also, and one has to be very innovative to bring their point across using time and space figures. For time and space plots, one can use 3D animation features in Tecplot. The procedure to make animation or movies using 3D data is detailed in Appendix A.

To show the grid spacing and flow features around the building, every other points are plotted in Figure 5.22. Here, the planes J = 76 for x–z plot and K = 4 for x–y plot are considered. One can see that the flow goes around and over the building. The flow separations of windward and leeward sides of the building are noticed in Figure 5.22a. The flow separation on the windward edge is yet to be captured due to the grid limitation. The flow separation on the side of the building can be seen in Figure 5.22b. It is noticed that the size of the vortex changes from one plane to another from animation of each plane. Hence, it is a complex flow. The close-up view of the plan and elevation of the building is shown in

(a)

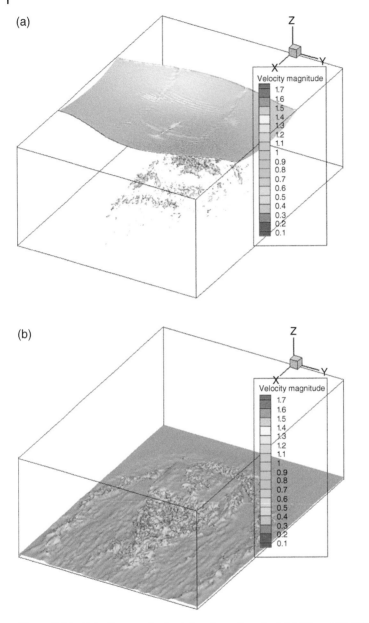

(b)

Figure 5.21 Velocity magnitude plot as isosurface for (a) 1.35 and (b) 0.31.

Figure 5.23. Here, all the points are considered. To show the effect of low pressure on velocity, in Figure 5.24, both are shown as close-up along the centerline. One can see that at low-pressure region, vortices are formed. The major difference is here that the vortices are 3D, and hence the shape will be varying from one slice to another.

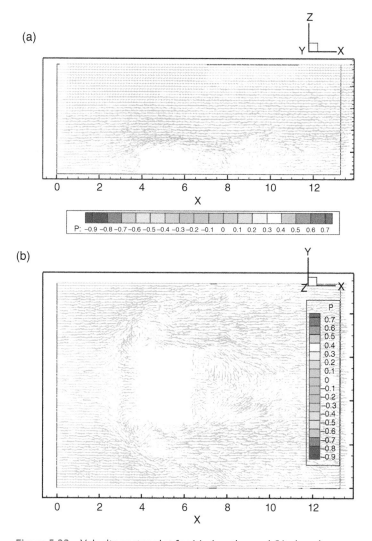

Figure 5.22   Velocity vector plot for (a) elevation and (b) plan view.

5.5.14   File Detail: prcon.plt

This file writes pressure on the building:

```
c.....PRESSURE CONTOUR-tecplot
      write(3,*)'VARIABLES = "X","Y","PM","PMAX","PMIN"'
      write(3,*)'ZONE I=',IML, ',J=',JML,',F=POINT'
      do j=1,jml
      do i=1,iml
      write(3,*)xl(i),yl(j),(pp(i,j,K),K=2,4)
      end do
      end do
```

(a)

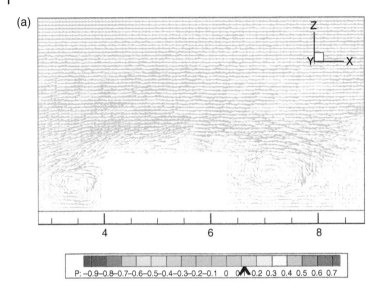

P: −0.9 −0.8 −0.7 −0.6 −0.5 −0.4 −0.3 −0.2 −0.1 0 0.1 0.2 0.3 0.4 0.5 0.6 0.7

(b)

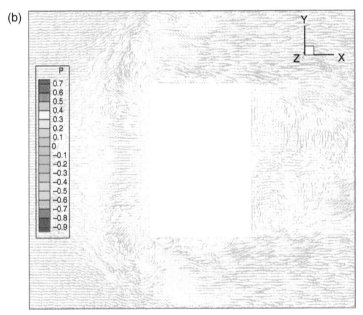

Figure 5.23   Zoomed velocity vector plot for (a) elevation and (b) plan view.

The pressure coefficient C_p on the roof and sidewalls are plotted in the xy plane to have a good view. The details of converting 3D image to 2D plane are shown in Figure 5.25. The face 1 is the windward side of the building. The corner regions in the 2D plane are kept at a big number so that they can be filtered from our plot using coloring option in Tecplot. The C_p mean, minimum, and maximum are plotted in Figure 5.26.

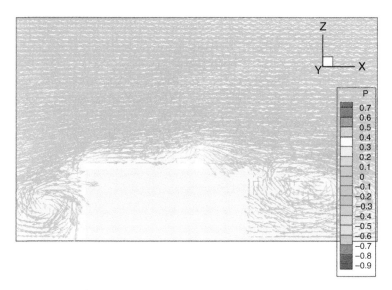

Figure 5.24   Close-up view of velocity vector plot with pressure contour.

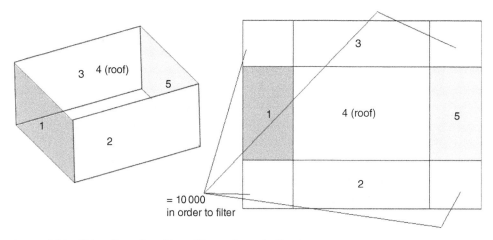

Figure 5.25   Sidewalls and roof from 3D are shown in 2D plane.

For specific lines, one can also retrieve the data. To retrieve data to plot pressure on the roof, use **gdata.f** and change the following info in the program for the given grid:

kh = 17,   ib1 = 65,   ib2 = 101,   jb1 = 49,   jb2 = 103

The program reads data from prcon.plt and writes the file: **pcent.plt.** The data from pcent.plt is copied and pasted into pcent-02-wind-v2.plt.

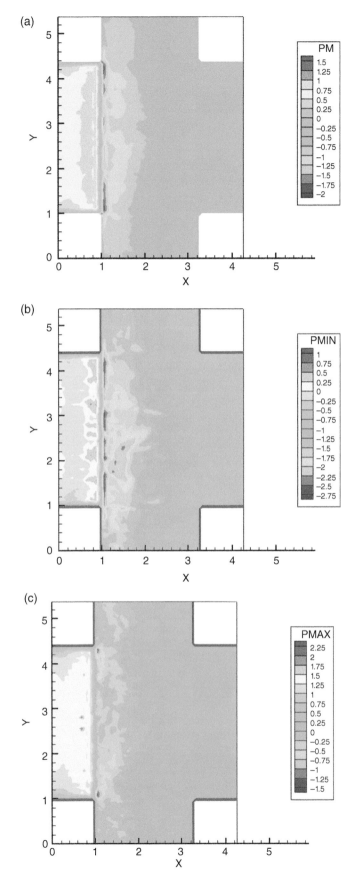

Figure 5.26 Pressure coefficient C_p on sidewalls and roof. (a) Mean, (b) minimum and (c) maximum.

5.6   Unsteady Flow over Building

5.6.1   Pressure on the TTU Building with Inflow Turbulence

The peak pressure is increased extensively when the flow is turbulent. As an example, the peak pressure coefficients reported by WT and field measurement for the TTU building in Figure 5.1b are −8 and −17, respectively. Whereas from CFD without inflow from Section 5.5.11 is less than −2.5. Even though CFD and measurement values are not one to one comparison, this gives an idea on the range. The CFD values may tend toward WT value when the inflow turbulence is included as seen in the following sections.

Turbulence is a 3D phenomenon, because in turbulent flow the eddy gets stretched and twisted. Hence, one has to perform 3D-CFD modeling. This phenomenon does not happen in a 2D modeling. The complexity of computing turbulent flow comes in due to the phenomena of energy cascade. During energy cascade, the low-frequency waves are split into high-frequency waves. If proper refinements are not provided, these high-frequency waves can produce numerical instability as explained by Phillips (1959). By using LES modeling, these high-frequency waves can be dissipated by the turbulent eddy viscosities. A lot of fundamental research is happening even now to understand their issues.

In the CFD modeling, turbulence is considered using LES in the recent years. There are several methods available to generate turbulence as initial conditions and BCs. Brief introduction is given in the next section.

5.6.2   Inflow Turbulence Generation Methods

The main methods used in inflow generation as discussed in Aboshosha et al. (2015) are

  i) Precursor database method
 ii) Recycling method
iii) Synthetic turbulence method

Precursor Database Method: In the precursor database method, turbulence is generated inside the computational domain as a channel flow first. Then the generated flow is used as initial condition or inflow to study the turbulent effect on building. Murakami et al. (1987) used the method of initial condition. First, they generated the turbulent flow without building, and then they introduced the building in the same flow. Lim et al. (2009) used the second approach.

Recycling Method: Lund et al. (1998) introduced the recycling method where the flow at the outlet is rescaled and introduced in the inflow. This method is used in wind engineering by Nozawa and Tamura (2002) and Kataokoa and Mizuno (2002).

The earlier mentioned two methods are not that attractive because of more computational time. In both methods, the largest length scale in the inflow is the length of the computational domain of the initial flow. The other issue is the loss of energy in turbulence due to energy cascade. This issue is not addressed very much in the past. For some understanding, one can refer to Selvam et al. (2020). Liu and Pletcher (2006) reported a review on precursor database and recycling method.

Synthetic Turbulence Method: The synthetic turbulence method has been used extensively in the last few years, and further developments are happening. They are far more efficient computationally and they have better control on turbulence properties compared to field observations. Several methods are developed under this category. The major methods currently in use are:

1) Random Fourier method (RFM)
2) Digital filter method (DFM)
3) Synthetic eddy method (SEM) and divergence-free synthetic eddy method (DFSEM)

Random Fourier method (RFM): In the RFM, Fourier series are used to generate the inflow turbulence. Aboshosha et al. (2015) provide a good review on this topic. Smirnov et al. (2001) provide the basis for generating the inflow for a particular turbulence. The divergence-free condition is satisfied using the method of Kraichnan (1970). Smirnov et al. (2001) showed that mass conservation or divergence-free condition is satisfied theoretically by the inflow using random flow generator (RFG) method for isotropic turbulence and nearly divergence-free for anisotropic turbulence. Hence, there is some error in the continuity equation. Further details on derivation of Smirnov et al. (2001) work and continuity error in that work are provided in Yu and Bai (2014). Modification to eliminate the continuity error is provided in Yu and Bai (2014).

Improvements were made by Huang et al. (2010) for the RFG method to suit for atmospheric boundary layer (ABL) flow. This is called discretizing and synthesizing random flow generation (DSRFG) method and here they used proper turbulence spectrum in three directions. They used three equations to satisfy the mass conservation, and they say the proposed procedure satisfies the mass conservation theoretically. The claim is valid for isotropic turbulence with constant turbulence length scale. When the length scale (Li) used is different in each direction, the proposed method of satisfying mass conservation has high error as explained as follows. In their work, they used constant Li in each direction or in the ith direction.

Let us write the turbulent velocity U_i as a sum of sine function as reported in Yu et al. (2018) at direction x_j as:

$$U_i(x_j, t) = \sum p_{i,n} \sin\left(k_{j,n} y_{j,n} + 2\pi f_n t + \psi_n\right) \quad \text{for } n = 1, N$$

Here, $p_{i,n} = \sqrt{(2S_{u,i}(f_n)\Delta f)}$

The definitions of each variable as reported in Yu et al. (2018) are

U_i velocity in three directions i = 1–3
$S_{u,i}$ wind spectrum in the i direction at frequency f_n
Δf width of the frequency in the wind spectrum
ψ_n uniform random distribution between 0 and 2π
$y_{j,n}$ $x_{j,n}/L_{j,n}$ where $L_{j,n} = U_{ave}/f_n c_j r_j$ is the length scale for a given frequency f_n
x_j coordinate for j = 1–3

For further details, one can refer to Yu et al. (2018). To satisfy the continuity equation, one has to satisfy the following condition:

$$\partial U/\partial x + \partial U/\partial y + \partial W/\partial z = 0.0$$

Differentiating each term gives the following condition for each frequency to satisfy continuity:

$$p_{n1}k_{n1}/L_1 + p_{n2}k_{n2}/L_2 + p_{n3}k_{n3}/L_3 = 0.0$$

In addition, they use the relation $k_{n1}^2 + k_{n2}^2 + k_{n3}^2 = 1$

The earlier mentioned equations satisfy the divergence condition theoretically. Huang et al. (2010) do not divide the first two equations by L_i, and hence they violate divergence-free condition to some level. It is suspected that they assumed L_i to be same for anisotropic flows and canceled it.

Aboshosha et al. (2015) modified further to have coherence for the velocity and it is called Consistent Discrete Random Flow Generator (CDRFG) method. They included a MATLAB code for others to use. They used the same procedure as Huang et al. (2010) to satisfy the continuity equation, and hence similar error persists. In addition to Lj different in each direction, they also depend on mean velocity and frequency. The mean velocity changes with respect to height, and this effect is not considered in satisfying the continuity equation.

Yu et al. (2018) satisfy the equation theoretically, and their method is called narrow band synthesis random flow generator (NSRFG). Only issue is that L_i is a function of velocity and hence changes in each direction as well as with respect to height for each frequency. They consider the change in x, y, and z directions of L_i. The height effect introduces some error if it is considered or not considered. In this chapter, **we will use NSRFG method as inflow**.

In the program **yif2.f**, the turbulence spectrum considered is a simpler one reported in Section 2.10. The wind spectrum is a function of only f. For more sophisticated wind spectrum, one can refer to Aboshosha et al. (2015). In that model, turbulent intensity and turbulent length scale are also considered as input.

All the methods have some form of continuity equation error. Some methods have more error than properly used RFM. The RFM and DFSEM have the advantages of controlling the maximum input frequency to some level. The DFM does not have provision to control the input frequencies and also it has very high continuity equation error as reported in Mansouri and Selvam (2020).

5.6.3   Inflow Turbulence Effect on Flow and Pressure Without Building

For this study, the flow details reported in Moravej (2018) for 1:6 scale WT study of the TTU building are considered. From the WT spectrum, we can get $f_{max} > 10$ and $f_{min} = 0.07$, where f_{max} and f_{min} are the nondimensional frequency from WT measurements. Here, f_{min} is taken just the left of peak point on p59 of Moravej thesis.

In the LES computation, one has to pay attention to the maximum frequency used in the inflow turbulence. As discussed in Chapter 2 in the LES modeling, the frequency spectrum considered in the CFD computation without any modeling is up to f_{LES} as shown in Figure 5.27, and this part is captured accurately by CFD. When $f_{max} > f_{LES}$, this part is modeled by LES modeling equations with less accuracy. The f_{LES} can be determined from the grid spacing used in the inflow region or the largest grid spacing used in the computational domain. The largest frequency that can be captured by the finite difference method (FDM) grid is discussed in Chapter 3 using the wave equation modeling. From that investigation, we can say that the minimum wavelength should be 4h and the corresponding frequency

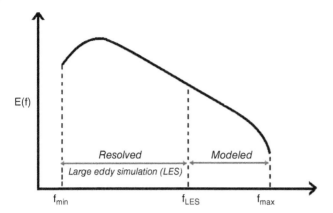

Figure 5.27 Wind spectrum with respect to LES.

$f_{LES} = f_{grid} = f_{max} = vH/4h$ where h is the largest grid spacing used in the flow. Since we plan to use constant grid spacing, then it is simply 4h. For h = H/16 grid, the f_{LES} comes to be 4. If the f_{max} used in the inflow computation is greater than f_{LES}, then due to numerical error and continuity equation error, spurious or unrealistic pressures are produced in the computation. Mansouri et al. (2020) discuss this in detail. In Mansouri et al. (2020), the continuity equation error at the inlet is calculated using FDM approximation of continuity equation by taking three time slices. To perform this calculation, specific program has to be written in our research. In the commercial or open-source code like OpenFOAM, it is difficult to perform this. In the work of Atencio (2021), the velocity and pressure are plotted in time at the building location without building and from that she could evaluate as shown in Figure 5.28. This method makes it easier to implement in the regular code. Here, $f_{min} =$ 0.07 and f_{max} is kept as 4.0 and 10.0 for H/16 grid. From Figure 5.28b, one can see the spurious or discontinuous pressure at the building location without building for $f_{max} = 10$. This effect is far less for $f_{max} = f_{grid} = 4$ as shown in Figure 5.28a. One can also see how FDM filter the high-frequency velocity at the building location to that of the velocity at the inflow.

So far, we discussed for a given grid spacing h, what f_{LES} one should use. To get a good accuracy what is the h one should use and how much part of the spectrum one should consider in the CFD modeling is not clearly investigated so far. Mansouri et al. (2020) considered H/8, H/16, and H/24 grid spacing for the TTU building. They used CDRFG method of inflow turbulence generation reported in Aboshosha et al. (2015). The corresponding f_{LES} for the smallest wavelength of 4h comes to be 2, 4, and 6. They computed the peak pressure for the three cases and reported that there is a steady improvement in the peak pressure results. The error is very high for $f_{max} = 2$ when compared to 1 : 6 WT measurement. The H/16 grid spacing is in reasonable comparison with WT measurement for roof pressures, but the H/24 grid pressures are much better. The H/24 grid spacing has more than eight million grid points. One could see the convergence with grid refinement using this approach. One could conclude that H/16 grid spacing is reasonable for preliminary investigation.

(a)

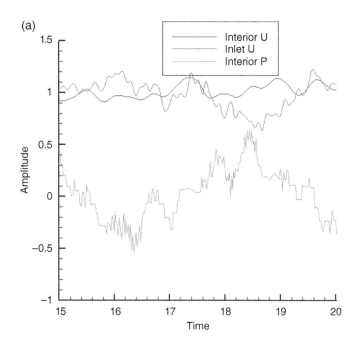

(b)

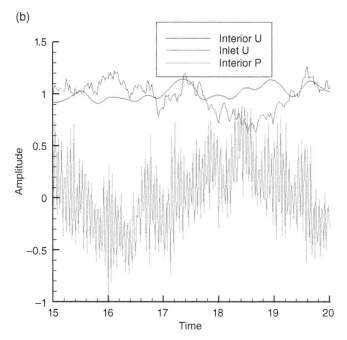

Figure 5.28 Spurious pressure reduction: (a) $f_{max} = 4$ and (b) $f_{max} = 10$.

For h = H/16 grid, the grid frequency f_{grid} = H/4h = 4 and hence f_{max} = f_2 = 4 and f_{min} = f_1 = 0.07 will be used in the computer program. The reference velocity U_{ave} (UREF) and height H (HREF) are taken as 8.6 m/s and 4 m, respectively.

Then the actual frequency n_1 and n_2 are calculated for f_1 = 0.07 and f_2 = 4 as:

$$f_1 = n_1 H/U, \quad n_1 = f_1 U_{ave}/H = 0.07(8.6/4) = 0.15 \text{ Hz}$$
$$f_2 = n_2 H/U, \quad n_2 = f_2 U_{ave}/H = 4(8.6/4) = 8.6 \text{ Hz}$$

Then the input DF, IFS, and IFE are calculated from the relationship given in Section 5.5.7 by assuming DF. DF depends on the number of frequencies considered in the sine series:

$$n_1 = DF(2IFS - 1) \text{ and } n_2 = DF(2IFE - 1)$$
$$n_1 = DF(2IFS - 1) \text{ and } n_2 = DF(2IFE - 1)$$
$$IFS = (n_1/DF + 1)/2 \text{ and } IFE = (n_2/DF + 1)/2$$

Here, IFS and IFE have to be whole numbers for input.

For f_1 = 0.07 and f_2 = 4, n_1 = 0.15 Hz and n_2 = 8.6 Hz, the corresponding DF, IFS, and IFE are

If DF = 0.1 Hz, IFS = 1 and IFE = 43, n_1 = 0.1 Hz (f_1 = 0.047), and n_2 = 8.5 Hz (f_2 = 3.95).
If DF = 0.2 Hz, IFS = 1 and IFE = 22, n_1 = 0.2 Hz (f_1 = 0.093), and n_2 = 8.6 Hz (f_2 = 4).
For the current computation, let us use DF = 0.1 Hz, IFS = 1, and IFE = 43.

5.6.4 Computation of Wind Spectrum Using the Program yif2.f

Time to Run: To calculate the wind spectrum from CFD and compare with WT spectrum, one needs to have sufficient duration of data. This will be determined from the minimum frequency f_1. The smallest frequency f_1 considered in the flow is around 0.05, and hence the period of the largest eddy size is 20 time units. Hence, the CFD flow has to be conducted minimum 10 time units for flow to reach steady state plus 20 time units to capture one largest eddy. In the CFD computation, TTIME is taken as 30 time units.

Modification of Input Data for Flow Without Building and Considering Inflow Turbulence: We will modify the input from Section 5.5.7 by considering KH = 1 and INFLT = 1. The user manual is provided in the same section. The relevant numbers that are changed in the input data are shown in bold:.

Input data for the program-yif-i.txt
213,151,81,65,101,49,103,**1**,0.02,**30.0**
0.195,0.006,5000,4.E-7,13.25,9.375,5.,**0.1,1,43,4.0,8.6,1**

Using the earlier mentioned data, the **yif2.f** program has to be run and the output files will be processed for further analysis. Mostly, six subiterations for each time step are noticed. Whereas with building and without inflow, only four subiterations at each time step are noticed. The total computer time the program yif2.f took was little more than 11 hours in a single-processor Linux system

Using the **yif-o3.plt** file, first the inlet and building location U velocity and pressure at the building location are plotted as shown in Figure 5.29 to see the initial time to reach

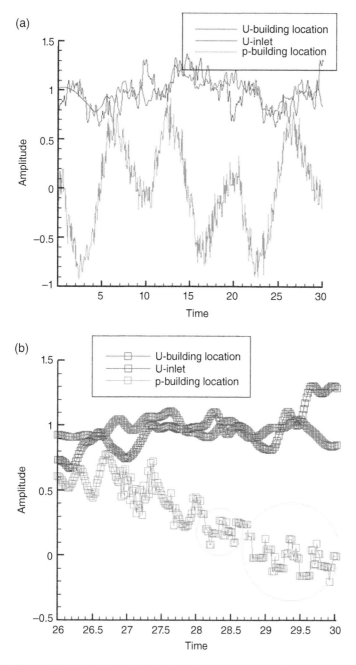

Figure 5.29 Variation of U velocity at inlet and building location and pressure at building: (a) for the entire CFD computation and (b) close-up view.

reasonable steady state and to evaluate the effect of continuity equation error. From Figure 5.29a, one can see that at the building location, the model took about 10 time units to reach steady state. This part will not be considered for spectral analysis. There is some noise in the pressure plot as shown in the close-up view in Figure 5.29b due to continuity equation error. Especially at the cyan color circled region, one can see the sudden changes because of this error. As discussed before, there is no inflow turbulence generation method available at this time to eliminate this error. The $f_{LES} = f_{max} = 4$ as input reduced the error extensively as illustrated in Figure 5.28b.

The **yip-o3.plt** has to be modified to consider the data from 10 to 30 time units for both at inlet and at building location. This can be done with using either Excel or a program **wcon2.f** listed as follows. After the modification, the **spd1.exe** program introduced in Chapter 2 will be used for wind spectrum calculation. The number of data points comes to be 1003 and the nfreq is considered to be less than half in running the **spd1.exe** program. To compare with the given Kaimal spectrum for U velocity from Chapter 2, the equation is modified for $U_{ave} = 8.6 \, \text{m/s}$, $H = 4 \, \text{m}$, $z_0 = 0.024 \, \text{m}$, and $u^* = 0.735$ as follows:

$$nSu/u^*2 = 200f/(1 + 50f)^{5/3}$$

$$Su = 200u^*2H/U_{ave}(1 + 50f)^{5/3} = 54.4/(1 + 50f)^{5/3}$$

The earlier mentioned expression is inserted in the **spd1.f** for comparison.

```
Program: wcon2.f
c       program wcon2.f, 10/18/2020
c       CALCULATE THE UAVE AND GET THE NECESSARY VELOCITY FOR SPECTRAL
c       ANALYSIS
c       CONVERT DIMENSIONAL VELOCITY TO NON-DIMENSIONAL VELOCITY
        PARAMETER(NS=10000)
        IMPLICIT REAL*8 (A-H,O-Z)
        DIMENSION V(NS,3)
        OPEN(1,FILE='yif-o3.plt')
        open(2,FILE='spd-i.txt')
        DT=0.02
        NP=1501
        H=1.0
        ID=499
        SUM=0.0
        IC=0
        DO I=1,NP
        READ(1,*)time,u1,u2,u3,V(I,1)
        IF(I.GE.ID)THEN
        IC=IC+1
        SUM=SUM+V(I,1)
        V(IC,2)=V(I,1)
```

```
      END IF
      END DO
      UAVE=SUM/IC
      DTN=DT*UAVE/H
      Print *,IC,uave,dtn
      TTIME=(IC-1)*DT
      NFREQ=IC/2-1
      WRITE(2,*)IC,TTIME,DT,NFREQ
c.....WRITE NON-DIMENSIONAL VELOCITY
      DO I=1,IC
      V(I,3)=(V(I,2)-UAVE)/UAVE
      T=(I-1)*DT
      WRITE(2,*)T,V(I,3)
      END DO
      STOP
      END
```

The peak and rms for the wind at the inlet are calculated as 0.37 and 0.17 units, respectively. Hence, turbulence intensity is 17%. The corresponding values at the building location come to be 0.29 and 0.11. The peak and rms values reduced at the building location due to numerical error as well as due to energy cascade. The wind spectrum for both at inlet and building location at building height is plotted in Figure 5.30. The $f_{max} = 4$ is the cutoff frequency provided at the inlet, and the plot in Figure 5.30a clearly shows that. The wind spectrum is parallel to the given Kaimal spectrum. At this time, 43 frequencies are considered in the input (IFS = 1 and IFE = 43) for calculating the wind spectrum. If the IFE would have been increased, the plot may be much closer to the Kaimal spectrum. The spikes in the inlet plot in Figure 5.30a may be the discrete frequencies considered by the program, and this spike is not seen at the building location. From Figure 5.30b, one can see that the energy decreases sharply around f = 3 at the building location because of numerical error in approximating the f = 4 waves. For f = 3, the number of spacing to represent the smallest wavelength will be little more than 5h. The wind spectrum is much smoother in Figure 5.30b than Figure 5.30a.

5.6.5 Peak Pressure on TTU Building Using Inflow Turbulence

Input detail: The change to include the building is to give KH = 17 in Section 5.6.4. In addition, the TTIME is changed to 100 time units to consider longer time to calculate the pressure statistics.

The initial 10 time units are not considered for any of the statistics. For the DF, IFS, and IFE values used in the input, the nondimensional frequencies come to be $f_1 = 0.047$ and $f_2 = 3.95$. In this case, the longest wavelength in the computation is 21.3 units, and hence 4 wavelengths are included in the 90 time units. The CFD computation took 58 hours, and most of the maximum subiteration is around 6 to have convergence.

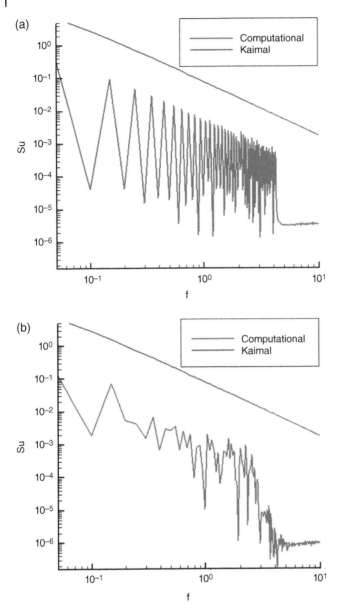

Figure 5.30 Wind spectrum: (a) at the inlet and (b) at the building location.

Pressure Variation in Time at Four Points on the Building

The pressure variation in time is plotted using **yif-o.plt** file at four points on the building as shown in Figure 5.31. Overall about 15 frequencies are noticed in the plot. The peak pressure coefficient C_p is close to −4 as noticed in Figure 5.31. The time variation of the pressures including inflow turbulence is much different from the one plotted before without inflow turbulence.

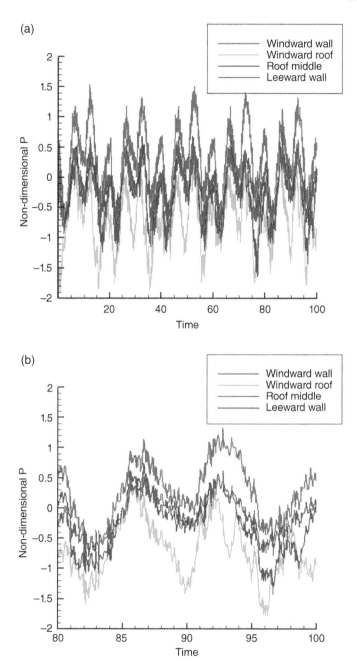

Figure 5.31 Pressure variation on building in time: (a) full view and (b) close-up.

Comparison of Pressure Coefficients Along the Centerline of the TTU Building

The mean, maximum, and minimum pressure coefficients are calculated from 10 time units to 100 time units and are written in the data file **yif-o2.plt**. The CFD values are plotted in Figure 5.32. The CFD mean C_p is compared with 1 : 6 (WT6) and 1 : 50 (WT50) scale WT measurements from Moravej (2018) in Figure 5.33a. Figure 5.33b compare the CFD minimum peak C_p with WT. From the figure, we can say that the mean values are slightly higher than WT6 values in the middle of the roof. The peak values are lower than WT6 on the windward edge. For the given range of nondimensional frequencies (0.047 and 3.95) in the input, the minimum peak pressure coefficients are far lower than 1 : 6 scale WT spectrum. The CFD peak C_{pmin} of -3.8 on the windward edge of the roof is lower than the WT6 value of -3 and much closer to the field measurement value of -4 as reported in Selvam (1997) and Moravej (2018). The corresponding C_{pmin} for no-inflow turbulence is -2.4. Hence, the inflow turbulence has a significant impact on the peak pressures. On the other hand, when $f_1 = 0.1$ was used by Selvam et al. (2020), Mansouri et al. (2020), and Atencio (2021), the C_p values were comparable to WT6 measurements as reported in Figure 5.34 for comparison. The $f_1 = 0.1$ may be the spectrum closer to WT spectrum. The $f_1 = 0.047$ tends toward field spectrum peak position of 0.02 or 0.03. More systematic study is needed to know which part of the Kaimal spectrum one needs to consider for WT comparison. The reason for this approach is that there is no analytical expression for WT spectrums for low-rise building study. The peak value of the wind spectrum from WT may be considered. This needs verification.

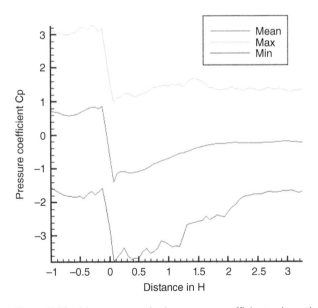

Figure 5.32 Mean, max, and min pressure coefficients along the centerline.

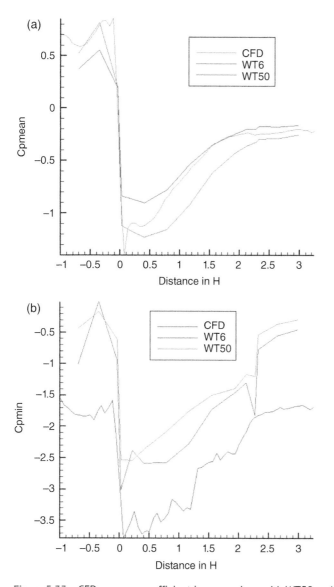

Figure 5.33 CFD pressure coefficient in comparison with WT50 and WT6 measurement: (a) C_p mean plot and (b) C_p min plot for $f_{min} = 0.05$ and $f_{max} = 4$.

On the windward and leeward wall, the CFD mean C_p are in good comparison with WT6 data. The minimum peak C_p is much lower than WT measurements. Similar trend is noticed in Mansouri et al. (2020) and Atencio (2021) CFD results also. This issue is yet to be resolved with respect to CFD inflow turbulence methods. May be the continuity equation error is much higher close to the ground.

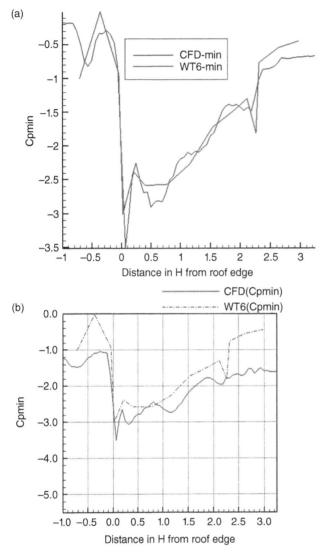

Figure 5.34 Peak C_p for $f_{min} = 0.1$. *Source:* (a) Taken from Selvam et al. (2020) and (b) taken from Mansouri et al. (2020).

Wind Spectrum at the Inlet

The wind spectrum at the inlet is plotted in Figure 5.35 by considering 5001 points (100 time units) available from **yif-o3.plt**. Here, the spikes are far more distinct than 20 time unit plot without building study, but they are much closer to the given spectrum. The corresponding velocity spectrum somewhere in the inlet may be of interest to observe.

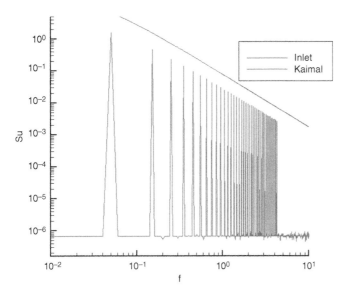

Figure 5.35   Velocity spectrum at the inlet using 100 time unit data.

Visualization of Velocity and Pressure After 100 Time Units

After converting **yif-p.plt** file into **yip-pe.plt** using **preplot** as we discussed before, few visualizations are presented for illustration. Since this the time-dependent flow and visualizing at one time step will not give a complete picture of what is going on, animation in 2D or 3D may help. Since each data file (**yif-p.plt**) is about 0.2 Gbytes at this time, no animation is made. Figure 5.36 shows the isosurface plot for pressure. The isosurface considered

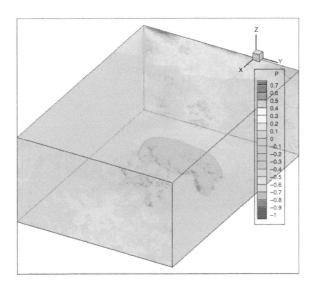

Figure 5.36   Isosurface plot.

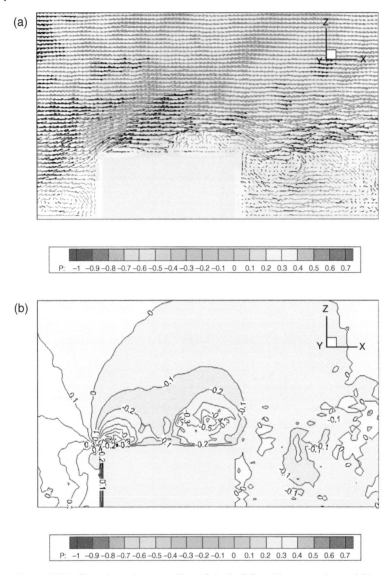

Figure 5.37 Plot along the centerline of the building (a) vector plot and (b) pressure plot.

is −0.15 units. From that one can see two major isosurfaces on the windward side and roof. The size of this will be changing in time. Figure 5.37 shows the pressure contour plot and vector plot along the centerline of the building. The vortices developed at the edge of the roof, and when they separated from the edge, the size grew and showed closed to the lee- ward part of the roof. These are the places 3D animation may help to see well. The vortices are three-dimensional in shape and they change in shape as time goes on. The vortices developed about z-axis are shown for K = 5 or at 4h from the ground in Figure 5.38.

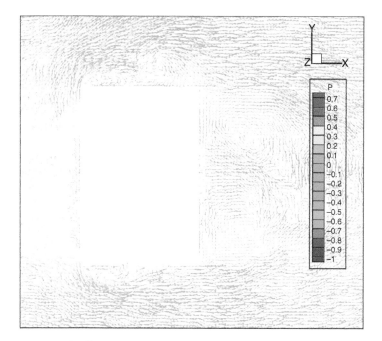

Figure 5.38 Plan view of the building at K = 5.

Figure 5.39 shows the pressure contours on the roof. One can see the peak negative pressures along the edge of the roof.

Analysis of Pressure Statistics for Flow Along the Shorter Side of the TTU Building

Using the file **prcon.plt** file, the mean and extreme pressure coefficients are visualized on the complete building in this section. The details of the faces are explained in Figure 5.25. The mean and peak C_p are calculated from 10 time units to 100 time units. Figure 5.40 shows the mean pressure coefficients C_p. Discuss the maximum value here to ASCE 7-16 mean values. Figure 5.41 shows the minimum and maximum peak pressure coefficients C_{pmin} and C_{pmax}.

The minimum average C_p on the roof from Figure 5.40 is −2.06. The minimum reported by the ASCE on the roof is −1.3 for MWRFS. Similarly, the smallest C_{pmin} reported on the roof is −4.7 from Figure 5.41a. There are several places the C_p is smaller than −4.5 as shown in Figure 5.41a. This value of −4.7 is smaller than −3.2 (the minimum C&C) from ASCE 7-16 as reported in Section 5.1. If a CFD study is conducted for 45 to 60 degree angle of incidence, the roof corner C_p will be much smaller as discussed in Section 5.1. Hence, ASCE 7-16 is mainly taken as minimum force for design.

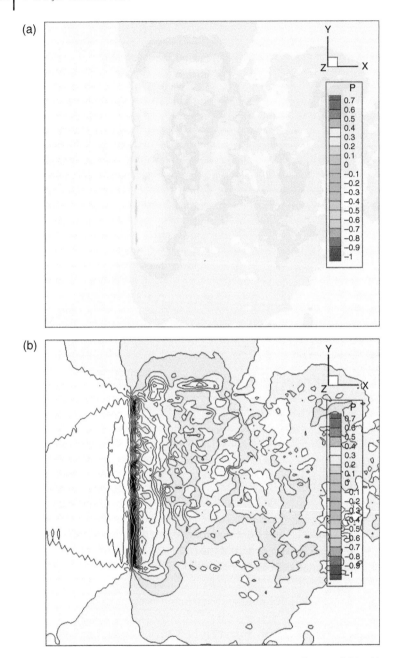

Figure 5.39 Pressure plot on the roof (a) without contour lines and (b) with contour lines.

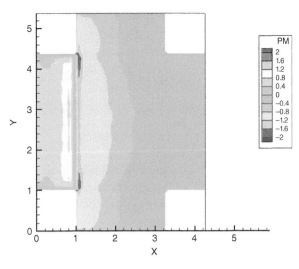

Figure 5.40   Mean pressure coefficients (C_{pmean}) on all the surfaces of the TTU building.

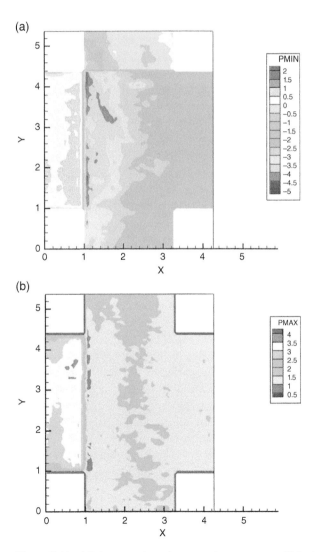

Figure 5.41   Minimum and maximum peak pressure coefficients on sidewalls and roof of the TTU building: (a) C_{pmin} and (b) C_{pmax}.

5.7 Flow Around a Cylinder and Practical Relevance to Bridge Aerodynamics

The flow around cylinder is a complex problem. The understanding of this problem will help to understand the issues in flow around bridges. The major difference between bluff cylinders and bridges is that the bridges are streamlined structure. As the wind speed increases or the Re increases, the flow slowly changes from laminar to complex flow behind the cylinder. For each structure, this transition depends on specific Re. In the turbulent wake region, vortex shedding starts to occur beyond this critical Re.

Due to vortex shedding, time-varying forces are created along (called drag force) as well as across the cylinder (called lift force) as discussed in Chapter 4. The amplitude of the time-varying forces in the drag direction is less than 10% of the mean value, and most of the time they are considered in design as well as code specification as some percentage of amplification to the mean loads. This phenomenon may be called buffeting. Whereas the lift force amplitudes can be as high as drag mean value as reported in Sumer and Fredsøe (2006), and hence the response of structures due to dynamic loading can be more than 10 times the static loading due to resonance. This ratio of dynamic deflection to static deflection is called dynamic load factor (DLF). The maximum DLF for periodic loading can be calculated as DLF $= 1/(2\xi)$, where ξ is the damping coefficient as discussed in Chapter 4. For practical problems, this value may be 0.1 or less. This amplification can lead to catastrophe unless proper precaution is taken in the design. For further wind engineering issues, please refer to Holmes (2007).

When fluid and structure interact, CFD and structural dynamics interact together. Then one has to be knowledgeable also in structural dynamics. Normally, one can calculate the time-varying force from CFD assuming that the structure is rigid and then use those forces on the structure and perform structural dynamic analysis. This is possible only if the maximum displacement of the structure is far less than the thickness of the structure in that direction. If the deformation is in the range of the thickness of the structure, then the fluid–structure interaction (FSI) has to be considered simultaneously. This is beyond this section, and we will consider it in the next chapter.

The flow over circular cylinder problem is far more complicated with respect to grid generation for CFD. The other issue is the location of the flow separation point. This point varies with respect to Re of the flow as discussed in Wu et al. (2004). This provides another challenge. As an example, for Re = 100, close-up view of the flow for a 41 × 61 grid is shown in Figure 5.42. The angle is around 114° as reported in Wu et al. (2004). Unless proper grid resolution is provided around the circular cylinder as well as in the radial direction, the location of the separation point cannot be captured accurately. If that is not done, the vortex shedding will not be captured accurately. So the proper grid resolution near the cylinder is essential. In the 41 × 61 grid, the tangential point varies by 6° and the location of a grid point is at 114°. The computed Cd = 1.365, amplitude of Cl = 0.25, and period of the vortex shedding T = 6.5 or St = 0.154 for Re = 100 are reported in Selvam

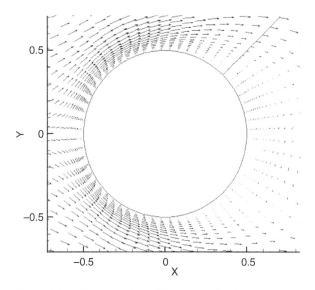

Figure 5.42 Close-up view of flow around a circular cylinder for a 41 × 61 grid.

(2020) for a 43×61 grid. For further details of others work, one can refer to Qu et al. (2013).

Because of the earlier mentioned grid generation challenge, we will consider flow around rectangular cylinders in this section for CFD modeling. In the case of square cylinder that is to be illustrated, the flow separation occurs at the windward edge of the cylinder and hence finer grid resolution close to the cylinder is not that essential. Due to square region, orthogonal grid is sufficient.

Flow Over Square Cylinder

The flow over square cylinder is investigated extensively as a benchmark problem. Two kinds of problems are considered depending on the top and bottom BCs. If the top and bottom boundaries are considered to be a wall as reported in Breuer et al. (2000), the blockage effect is considered. The inlet is a parabolic flow and the outlet is an open boundary. In the other type, the top and bottom boundaries are considered to be free stream, and the inlet has a constant velocity and outlet is an open boundary as reported in Gera et al. (2010), and a sketch is shown in Figure 5.43. Here, we will consider the later at this time.

We will use a grid generator program **rg.f** for grid generation. The input file is **rg-i.txt**. The output files are:

Rgrid.txt   text file that can be modified to use with **recyl3.f**
Rg-p.plt   plot file to see the grid

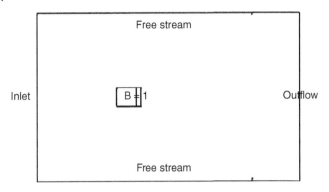

Figure 5.43   Flow over a square cylinder problem.

The input data in **rg-i.txt** details are:

```
READ(1,*)XLEN,YLEN,RMIN,RMAX,RMAY,HMAX,HBX,HBY,FAC
```

XLEN	XLENGTH OF THE CYLINDER
YLEN	YLENGTH OF THE CYLINDER
RMIN	MIN. SPACING
RMAX	MAX. DISTANCE FROM CENTER IN X
RMAY	MAX. DISTANCE FROM CENTER IN Y
HMAX	MAX. SPACING
HBX	SPACING ON THE CYLINDER IN X
HBY	SPACING ON THE CYLINDER IN Y
FAC	EXPONENTIAL FACTOR

In Figure 5.44, some of the input variables for **rg.exe** are explained.

Sample data file: 1.0,1.0,0.02,20.0,8.0,0.2,0.05,0.05,1.1

The **rgrid.txt** can be modified to reduce some of the unwanted data. The data is written in the following order in **rgrid.txt**:

```
WRITE(2,*)IM,JM,IMK1,IMK2,JMK1,JMK2
WRITE(2,20)(X(I),I=1,IM)
WRITE(2,20)(Y(J),J=1,JM)
20      FORMAT(5(E14.7,1X))
```

Here, IMK1 and IMK2 are the beginning and end node numbers of the cylinder in the I-direction, and JMK1 and JMK2 are the beginning and end node numbers in the J-direction.

The details of **recyl3.f** that is the NS solver for flow over square cylinder are:

The flow is solved with different solver. The momentum is solved implicitly by line iteration, and the pressure is solved by PCG and then u and v are updated. Maximum five

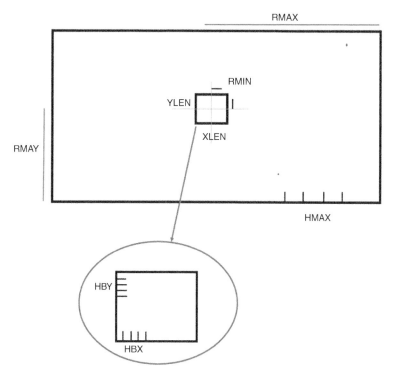

Figure 5.44 Flow over square cylinder grid generation definitions for **rg.f**.

subiterations are allotted to have the error reduced to specified accuracy. The maximum points in x and y allowed are 300.

The input file for the **recyl3.f** is **rcy-i.txt**.

The input data file **rcy-i.txt** is created by modifying the **rgrid.txt** file received as output file from the grid generator.

The detail of the input is:

```
READ(5,*)IM,JM,IMK1,IMK2,JMK1,JMK2,DTT,REN,TTIME
READ(5,*)(X(I),I=1,IM)
READ(5,*)(Y(J),J=1,JM)
```

IM & JM	number of points in x and y directions
IMK1,IMK2	beginning and end node number of the cylinder in the x direction
JMK1,JMK2	beginning and end node number of the cylinder in the y direction
DTT	initial time step. The program can calculate DTT after second time step
REN	Reynolds number
TTIME	total time units the program should run
X(I)	read x for I=1,IM
Y(J)	read y for J=1,JM

The outputs for this program are **recy-o.plt** and **recy-p.plt**.
recy-o.plt has time, Cd, and Cl for each time step
recy-p.plt has u, v, and p for each point to visualize.

First line of the sample data arrived for recyl3.f: 203, 133, 67, 87, 57, 77, 0.01, 100.0, 100.0

The flow considered is for Re = 100. The computational time to run for 100 time units is 8 minutes.

Computation of Drag and Lift Forces

The forces in the x and y directions are called grad and lift forces. They are defined as:

$$Fd = Cd\rho V^2 DL/2 \text{ and } Fl = Cl\rho V^2 DL/2$$

Here, Cd and Cl are called drag and lift coefficients, D is the width of the cylinder perpendicular to the flow, ρ is the density of the fluid, and L is the length in the z direction. They are calculated as follows:

$$Fd = \int[(\mu \partial Vt/\partial n)ny - Pnx]dS \text{ and } Fl = -\int[(\mu \partial Vt/\partial n)nx - Pny]dS$$

Here, nx and ny are the directional cosines of the normal with respect to x- and y-axis. The tangent vector has (ny,-nx) as the direction; refer Schafer paper.

In the **recyl3.f** program, Euler backward procedure is used. The program has QUICK and BTD schemes internally. This can be brought out in the input if needed. Also, provision has to be made for central difference. At this time, MCONV = 1 for BTD and MOCN = 2 for QUICK. The following plots use QUICK procedure for convection.

The vortex shedding started to occur around 10 time units. From Figure 5.45, we can see that the Cd mean comes to be 1.585 and Cl amplitude is 0.28. The period of the lift force oscillation is: 96.8 – 90.05 = 6.75. St = 1/T = 0.148. Sohankar et al. (1998) got Cdmean = 1.46, Cl-amplitude = 0.139 and St = 0.146.

Contour plots and vector plots are shown in Figure 5.46, and stream traces are shown in Figure 5.47.

5.8 Chapter Outcome

The pressures over the TTU building are computed and compared with WT and field measurements. When the inflow turbulence generator is introduced, the computed peak pressures are in comparison with WT and filed experiments. Hence, some reliability is gained from recent developments. The attractive feature of CFD is the speed at which one can get the pressure and velocity around buildings with very less cost. In a WT, it takes minimum several days to prepare the test sample and install the pressure taps and other measuring system in the required places. In addition, the cost involved in building the sample, conducting the experiment, paying for labor etc. are other challenges. Whereas using CFD, one can get results in few weeks and the cost may be less than $10K in the United States.

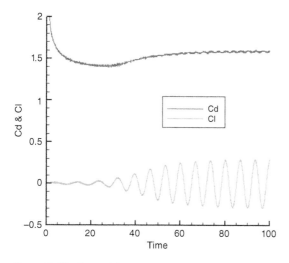

Figure 5.45   Drag (Cd) and lift (Cl) coefficient versus time.

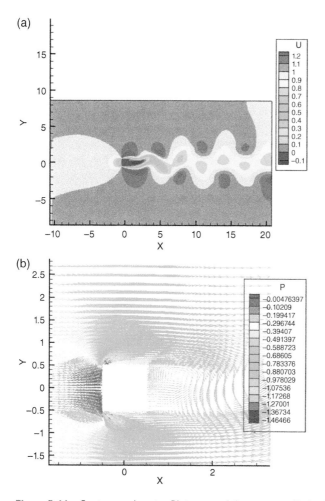

Figure 5.46   Contour and vector Plots around the square cylinder (a) contour plot and (b) vector plot.

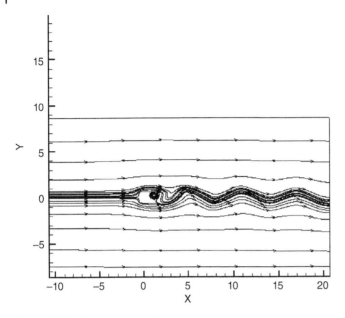

Figure 5.47   Stream traces plot.

The major challenge in the CFD is to validate the model with field and WT measurements. This is happening slowly as we discussed so far. Once reasonable reliability is achieved, one can use confidently CFD for practical applications.

The next problem considered is the flow over cylinder. This study should open the eye for bridge flow modeling. The main interest here is to illustrate the vortex shedding effects on structures.

The 3D-CFD work takes few days of computer run in single processor, and this can be reduced by order of magnitude by using parallel computing.

Problems

1 **Perform 2D-CFD** analysis for slope 45° (finer grid and larger domain than what is used in the illustration) and 26.56° (1 : 2 slope) escarpment. See with finer resolution you can capture flow separation at the bottom of the hill for 45°. Compare CFD velocities with ASCE 7-16 values.

2 Compute flow over square building using at least two different grids and compare the pressure around the building with ASCE. You can investigate the effect of XL, VIS, DT, NITERT, NITER, and RF for one of the grid.

3 **Problem for 3D building without inflow turbulence**: (i) Run for 20 time units using H/8 grid; (ii) plot the pressure coefficient along the centerline and compare with WT; and (iii) perform 2D and 3D visualization as done in the text using Tecplot or any other visualization software.

4 **Calculate vorticity** about z-axis or x-axis and visualize using isosurface and plot using Tecplot. For this, one needs to use "analyze" option discussed in Appendix A.

5 **Prepare 3D input for the Silsoe building** with details of how the input data is arrived. Details of the building and wind spectrum are provided in Mooneghi et al. (2016). For the frequency spectrum input detail DF, IFB, and IFE, consider both WT and field spectrum cases. The cases to be considered are (i) building without inflow turbulence, (ii) no building but with inflow turbulence, and (iii) building with inflow turbulence.

6 **Compute wind spectrum:** Using the 3D-CFD model with inflow turbulence generator, calculate the wind spectrum using H/8 grid (i) for WT spectrum and (ii) field spectrum.

7 **Problem for 3D building with inflow turbulence:** Using H/8 grid, compute the peak pressures for the TTU building and compare with H/16 grid reported in this chapter. Document your work with proper visualization.

8 **Problem for flow over square cylinder**: For Re = 100, compute the Cd and Cl plot and compare the results with available CFD and experimental results. Have proper visualization to make your points.

References

Aboshosha, H., Elshaer, A., Bitsuamlak, G.T., and El Damatty, A. (2015). Consistent inflow turbulence generator for LES evaluation of wind-induced responses for tall buildings. *Journal of Wind Engineering and Industrial Aerodynamics* 142: 198–216.

Atencio, Z. (2021). Frequency effect on peak pressure coefficients using the narrowband synthesis random flow generator (NSRFG) method. MSCE thesis, University of Arkansas.

Breuer, M., Bernsdorf, J., Zeiser, T., and Durst, F. (2000). Accurate computations of the laminar flow past a square cylinder based on two different methods: lattice-Boltzmann and finite-volume. *International Journal of Heat and Fluid Flow* 21: 186–196.

Caretto, L.S., Gosman, A.D., Patankar, S., and Spalding, D.B. (1972). Two calculation procedures for steady, three-dimensional flows with recirculation. In: *Proceedings of the 3rd International Conference on Numerical Methods Fluid Dynamices Paris*, Lecture Notes in Physics, Vol. 19, vol. II, 60–68. New York: Springer.

Chorin, A.J. (1968). Numerical solution of the Navier-Stokes equations. *Mathematics of Computation* 22: 745–762.

Gera, B., Sharma, P.K., and Singh, R.K. (2010). CFD analysis of 2D unsteady flow around a square cylinder. *International Journal of Applied Engineering Research* 1: 602–610.

Hirt, C.W. and Cook, J.L. (1972). The calculation of three-dimensional flows around structures and over rough terrain. *Journal of Computational Physics* 10: 324–340.

Hirt, C.W., Ramshaw, J.D., and Stein, L.R. (1978). Numerical simulation of three-dimensional flow past bluff bodies. *Computer Methods in Applied Mechanics and Engineering* 14: 93–124.

Holmes, J.D. (2007). *Wind Loading of Structures*, 2ee. New York: Taylor & Francis.

Huang, S.H., Li, Q., and Wu, J.R. (2010). A general inflow turbulence generator for large eddy simulation. *Journal of Wind Engineering and Industrial Aerodynamics* 98: 600–617.

Kataoka, H. and Mizuno, M. (2002). Numerical flow computation around aeroelastic 3D square cylinder using inflow turbulence. *Wind and Structures* 5: 379–392.

Kraichnan, R. (1970). Diffusion by a random velocity field. *Physics of Fluids* 13: 22–31.

Levitan, M.L., Mehta, K.C., Vann, W.P., and Holmes, J.D. (1991). Field measurements of pressures on the Texas tech building. *Journal of Wind Engineering and Industrial Aerodynamics* 38: 227–234.

Lim, H.C., Thomas, T.G., and Castro, I.P. (2009). Flow around a cube in a turbulent boundary layer: LES and experiment. *Journal of Wind Engineering and Industrial Aerodynamics* 97: 96–109.

Liu, K. and Pletcher, R. (2006). Inflow conditions for the large eddy simulation of turbulent boundary layers: a dynamic recycling procedure. *Journal of Computational Physics* 219: 1–6.

Lund, T.S., Wu, X., and Squires, K.D. (1998). Generation of turbulent inflow data for spatially-developing boundary layer simulations. *Journal of Computational Physics* 140: 233–258.

Mansouri, Z. and Selvam, R.P. (2020). Performance of different inflow turbulence methods provided in turbulence inflow tools for wind engineering applications. Report, Department of Civil Engineering, University of Arkansas.

Mansouri, Z., Selvam, R.P., and Chowdhury, A. (2020). Grid spacing effect on peak pressure computation on the TTU building using synthetic inflow turbulence generator. Report, Department of Civil Engineering, University of Arkansas.

Mooneghi, M.A., Irwin, P., and Chowdhury, A.G. (2016). Partial turbulence simulation method for predicting peak wind loads on small structures and building appurtenances. *Journal of Wind Engineering and Industrial Aerodynamics* 157: 47–62.

Moravej, M. (2018). Investigating scale effects on analytical methods of predicting peak wind loads on buildings. FIU Electronic Theses and Dissertations. 3799. https://digitalcommons.fiu.edu/etd/3799 (accessed 10 March 2022).

Murakami, S., Mochida, A., and Hibi, K. (1987). Three-dimensional numerical simulation of airflow around a cubic model by means of large eddy simulation. *Journal of Wind Engineering and Industrial Aerodynamics* 25: 291–305.

Nozawa, K. and Tamura, T. (2002). Large eddy simulation of the flow around a low-rise building in a rough wall turbulent boundary layer. *Journal of Wind Engineering and Industrial Aerodynamics* 90: 1151–1162.

Phillips, N.A. (1959). An example of non-linear computational instability. In: *The Atmosphere and the Sea in Motion*, 501–504. Rockefeller Inst.

Qu, L., Norberg, C., and Davidson, L. (2013). Quantitative numerical analysis of flow past a circular cylinder at Reynolds number between 50 and 200. *Journal of Fluids and Structures* 39: 347–370.

Richards, P.J. and Hoxey, R.P. (2012). Pressures on a cubic building-Part 1: Full-scale results. *Journal of Wind Engineering and Industrial Aerodynamics* 102: 72–86.

Richards, P.J., Hoxey, R.P., Connell, B.D., and Lander, D.P. (2007). Wind-tunnel modelling of the Silsoe Cube. *Journal of Wind Engineering and Industrial Aerodynamics* 95: 1384–1399.

Schlichting, H. (1968). *Boundary-layer Theory*. New York: McGraw-Hill.

Selvam, R.P. (1990). Computer simulation of wind load on a house. *Journal of Wind Engineering and Industrial Aerodynamics* 36: 1029–1036.

Selvam, R.P. (1992). Computation of pressures on Texas Tech Building. *Journal of Wind Engineering and Industrial Aerodynamics* 43: 1619–1627.

Selvam, R.P. (1996). Computation of flow around Texas Tech Building using k-ε and Kato-Launder k-ε Turbulence Model. *Engineering Structures* 18: 856–860.

Selvam, R.P. (1997). Computation of pressures on Texas Tech Buildingusing large eddysimulation. *Journal of Wind Engineering and Industrial Aerodynamics* 67 & 68: 647–657.

Selvam, R.P. (2010). Building and bridge aerodynamics using computational wind engineering. In: *Proceedings: International Workshop on Wind Engineering Research and Practice, May 28–29*. NC, USA: Chapel Hill.

Selvam, R.P. (2017). CFD as a tool for assessing wind loading. *The Bridge and Structural Engineer* 47 (4): 1–8. [Review paper-available as opensource].

Selvam, R.P. (2020). CFD Class notes. Department of Civil Engineering, University of Arkansas.

Selvam, R.P., Govindaswamy, S., and Bosch, H. (2002). Aeroelastic analysis of bridges using FEM and moving grids. *Wind & Structures* 5: 257–266.

Selvam, R.P., Chowdhury, A., Irwin, P., Mansouri, Z., and Moravej, M. (2020). CFD peak pressures on TTU building using continuity satisfied dominant waves (CSDW) method as inflow turbulence generator. Report, Department of Civil Engineering, University of Arkansas.

Smirnov, A., Shi, S., and Celik, I. (2001). Random flow generation technique for large eddy simulations and particle-dynamics modeling. *Journal of Fluids Engineering* 123: 359–371.

Sohankar, A., Norberg, C., and Davidson, L. (1998). Low-Reynolds-number flow around a square cylinder at incidence: study of blockage, onset of vortex shedding and outlet boundary condition. *International Journal for Numerical Methods in Fluids* 26: 39–56.

Strasser, M.N., Yousef, M.A.A., and Selvam, R.P. (2016). Defining the vortex loading period and application to assess dynamic amplification of tornado-like wind loading. *Journal of Fluids and Structures* 63: 188–209.

Sumer, B.M. and Fredsøe, J. (2006). *Hydrodynamics Around Cylindrical Structures*. New Jersey: World Scientific Publishers.

Vasilic-Melling, D. (1976). Three-dimensional turbulent flow past rectangular bluff bodies. PhD thesis, Imperial College, London. https://spiral.imperial.ac.uk/handle/10044/1/22815 (accessed 10 March 2022).

Versteeg, H.K. and Malalasekera, W. (2007). *An Introduction to Computational Fluid Dynamics*, 2ee. Prentice Hall.

Wu, M., Wen, C., Yen, R. et al. (2004). Experimental and numerical study of the separation angle for flow around a circular cylinder at low Reynolds number. *Journal of Fluid Mechanics* 515: 233–260.

Yu, R. and Bai, X.S. (2014). A fully divergence-free method for generation of inhomogeneous and anisotropic turbulence with large spatial variation. *Journal of Computational Physics* 256: 234–253.

Yu, Y., Yang, Y., and Xie, Z. (2018). A new inflow turbulence generator for large eddy simulation evaluation of wind effects on a standard high-rise building. *Building and Environment* 138: 300–313.

6

Advanced Topics

In the previous chapters, computational fluid dynamics (CFD) for wind engineering is introduced using flow around rectangular regions in two dimension (2D) and three dimension (3D). These simple geometries were taken for students to have hands on experience with CFD. Real problems have complicated geometry and hence needs sophisticated grid generators and takes more computer storage and time. In addition, only flow around building in ABL is considered. There are several topics in wind engineering where CFD is applied. Here an attempt is taken to discuss some of the topics with some detail.

In the past, Selvam (2008, 2010, 2017) reviewed the developments in CFD. These are good starting point. Recent reviews by Blocken (2014, 2018) have several references of his previous reviews, and these are also valuable information. We will refer more relevant references in each section.

Of all the different CFD applications in wind engineering considered here, topics such as pedestrian wind or wind environment around buildings, natural ventilation, and wind farm citing are well validated, and CFD methods can be used with confidence. Flow around bridges and computing critical velocity for bridge flutter are reasonably validated and some companies and agencies use the CFD model before going for wind tunnel (WT) testing. The mean pressures around building can be computed with reasonable confidence using CFD. The extreme pressure calculations over building are still evolving. Application of CFD for pollutant transport around building or street canyon and in complex terrain are developed well. Tornado and thunderstorm forces on building are still under research development.

6.1 Grid Generation for Practical Applications

In the previous chapters, simple rectangular grids were used for learning purposes. This reduced the time involved in generating the grid. This type of grid is called structured grid as discussed in Chapter 3. Under structured grid, body-fitted nonorthogonal grids are available for complex shapes as discussed by Thompson et al. (1985). In the same Chapter 3, unstructured grid concepts were introduced. For practical applications, structured, unstructured, and hybrid grids are used depending upon the problem. Some introductory discussions are available from Ferziger and Peric (2002) and Ho-Le (1988).

Computational Fluid Dynamics for Wind Engineering, First Edition. R. Panneer Selvam.
© 2022 John Wiley & Sons Ltd. Published 2022 by John Wiley & Sons Ltd.

CFD for practical application faces with complex shapes of buildings and bridges. Grid generation is a major challenge. For many practical applications, grid generation part of the project can take several weeks. Proper grid resolution in the region of flow separation is needed to get meaningful results. Without proper grid resolution, the computed pressure coefficients may not even be close to the realistic values. There are several sophisticated grid generators are available in the market. The following grid resolution concept is available in the literature:

1) Multilevel grid resolution and block-structured grid
2) Triangular grid for 2D and tetrahedra grid for 3D. There are several different grid generation concepts such as advancing front method and Delaunay triangulation method are available. For more discussion on different methods available, one can refer to Lo (2015).
3) Overlapping grids

Grid generated for flow around complex buildings using OpenFOAM snappyHexMesh is shown in Figure 6.1, and the corresponding velocity contour plot in Figure 6.2. This shows multiple levels of grid spacing. One coarse element is divided into two finer elements in this illustration. Other ratios are possible but for numerical accuracy and convergence this may be the better approach. More discussions will be done in Chapter 7 when OpenFOAM is introduced. By this process, one can reduce the number of grid points less than 1% of the finer grid.

6.1.1 Flow Around Complex Building and Bridge Shapes

For this type of application one needs unstructured or nonorthogonal grids. Proper grid generator is essential, and the Re effect necessitates find grid resolution at critical regions.

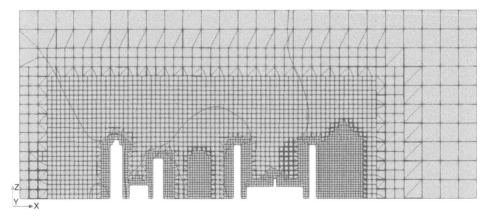

Figure 6.1 Grid generation for flow around complex building using snappyHexMesh concept in openFOAM.

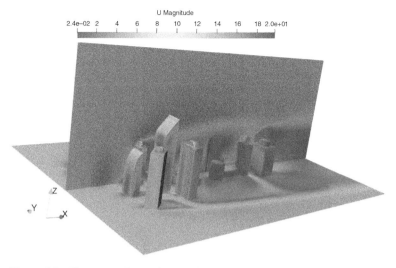

Figure 6.2 Flow around complex building using OpenFOAM: velocity contour plot.

6.2 Structural Aeroelasticity and Structural Dynamics

Aeroelasticity is defined as the study of the response of structures due to the interaction of inertial (structural dynamics), elastic (solid mechanics), and aerodynamic forces (CFD-fluid mechanics) on a flexible structure. The range of deformation (y) with respect to thickness (D) of the structure or y/D ratio may vary from 1 to 30 depending the type of phenomena as reviewed by Selvam (2017). Parkinson (1989) is a good review for galloping and vortex-induced vibration (VIV). The study of considering inertial and elastic properties of the structure in addition to CFD is called fluid–structure interaction (FSI). Davenport (1995) classifies aeroelastic interaction of wind on structures into three categories:

1) Forces created on structure due to on coming turbulence. In this case, it can be along wind and across wind responses. This can produce resonant responses
2) Forces created by vortex shedding behind the structure. The resonant response is mainly in the across wind direction. Usually the range of y/D ratio may be less than one as reported in Selvam (2017).
3) Forces created by motion of the structure. The response is mainly due to negative damping. This will occur as across wind response. The motion of the structure is many times more than the thickness of structure. This can cause aerodynamic instability. The topic of negative damping due to FSI is illustrated with simple problem in Chapter 4. The negative damping can occur only beyond certain critical velocity during FSI.

The response of the preceding three categories is illustrated by Davenport and Novak (1976) with a simple diagram, as shown in Figure 6.3. The sketch is not to scale. The aerodynamic damping occurs during VIV, galloping, and flutter. The aerodynamic damping may be positive or negative depending upon the range of free stream velocity, shape,

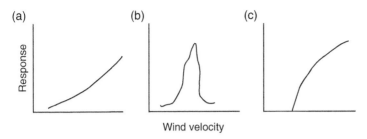

Figure 6.3 Aeroelastic interaction of wind on structures (a) response due to on coming turbulence, (b) response due to vortex shedding, and (c) response due to motion of the structure or aerodynamic instability.

and elastic properties of the structure. The total damping (ξ) is the sum of structural damping (ξ_s) and aerodynamic damping (ξ_a) as written later:

$$\xi = \xi_s + \xi_a$$

In the case of galloping and flutter, beyond the critical velocity the aerodynamic damping becomes negative damping. Often the negative damping may be less than structural or positive damping in the case of VIV. The aerodynamic damping can be measured from the WT test for a tall building as reported in Cao et al. (2012) for along wind response and Cheng et al. (2002) for across wind response. The aerodynamic damping is mostly positive for along wind vibration as reported in Zhang et al. (2015), and the range reported in that reference is 0.1–2.7%. For cross-wind response, the range is reported to be from 0 to −0.7% for tall building in the same reference.

Vortex shedding, galloping, and flutter are the major phenomena encountered under structural engineering. The topic of vortex shedding is illustrated in Chapter 5 for flow over square and circular cylinder using CFD by considering the structure rigid. Some discussion on flutter is done in Chapter 4.

6.2.1 Fluid–Structure Interaction (FSI) Methods

A detailed discussion on FSI methods is discussed by Selvam (2017). The FSI can be conducted by any one of the methods discussed later:

1) Use WT or CFD to get the fluids effect on structure. Using that information perform structural dynamics on the structure. This may be less computing, but the study may be linear analysis. This approach is used in the beginning when computational storage and time were limited for VIV, galloping, and flutter. Here get the force coefficients for various angle of attack and then use them in the dynamic analysis. An example for galloping of square section is discussed in Selvam (2017). Zhang et al. (2015) use this approach for Commonwealth Advisory Aeronautical Council (CAARC) tall building study. The discretizing and synthesizing random flow generation (DSRFG) inflow turbulence generator as inflow and detached eddy simulation to consider turbulence in CFD were used in that work.

2) Forced motion method for FSI: Here the structure is oscillated with known frequencies, and the response of the structure is measured. This approach is explained in detail in Simiu and Scanlan (1986) for bridge aerodynamics using WT measurements. These are called flutter derivatives. Then using these flutter derivatives, analytically the critical velocity for flutter is calculated. Walther (1994) and Larsen and Walther (1998) used this approach for bridge flutter calculation using CFD. Walther used 13 computer runs to calculate the flutter velocity of a bridge. Whereas using free motion method discussed in step 3, Selvam et al. (2001, 2002) used 5–7 computer runs to calculate the flutter velocity. Le Maitre et al. (2003) explain the implementation further. Larsen and Larose (2015) reviewed the method for cable stayed bridges.

3) Free motion method for FSI: Use WT or CFD and structural response or deformation together. Here at each time step the wind forces are calculated using CFD and then these forces are applied on the structure. Due to this force the structure moves and this movement is provided for the next CFD calculations. The detail of the implementation is explained in Selvam et al. (2001, 2002) for bridge flutter calculations.

In step 1, the mean force and moment coefficients are calculated for different angle of attack. These values are used to study the time-varying response of the structure. This method is applied for VIV and galloping study as reported in Parkinson (1989) and Selvam (2017) but not for flutter calculations. The method proposed in Step 2 and Step 3 can be used for galloping and flutter studies. For both methods, moving grid is needed, and they need special attention as explained in Selvam et al. (2002). Of the two methods discussed for flutter calculation, the free motion method is used extensively in the recent times and also the most economical one. This approach considers the nonlinear effect of the FSI.

6.2.2 Moving Grid for FSI Computation

The structure is formulated using Lagrangian coordinate system and the fluid is formulated using Eulerian coordinate system. When wind is moving due to the structural movement, then the effect of structural movement on fluid is handled via moving grid system. Selvam et al. (2002) used arbitrary Lagrangian–Eulerian (ALE) formulation with a special type of moving grid to perform the calculation efficiently. In the ALE-type moving grid if proper care is not taken to satisfy the geometric conservation law one may have to use very small-time step as discussed in Selvam et al. (2002) and the reference there in. Since only pitching and heaving motion is considered, a rigid body movement is considered for the whole fluid grid to satisfy the conservation law accurately. This reduced the computer time extensively. There are several methods available in the literature due to recent developments as reviewed by Miller et al. (2014). Important methods are listed later:

1) ALE formulation: To take care of the deformation of the grid around the moving grid, separate grid movement solver has to be used to move the grid at each time step. For more details, please refer to Selvam et al. (2002) and Miller et al. (2014) for key references.

2) Dynamic mesh method: In this approach at each time after the structure is moved to a new position, new grid is generated around the body. This needs an efficient grid generator and interpolation of the values from one grid to next. Mittal and Tezduyar (1992) and de Sampaio et al. (1993) used this for VIV of a circular cylinder.

3) Complete Eulerian approach: Here interface capturing method needs to be used to track the solid and fluid. This is similar to the work used by Selvam et al. (2006) for liquid droplet impacting a vapor bubble reported in Figure 1.2 in Chapter 1. Under this immersed boundary method and related method are reported in Miller et al. (2014)

4) Overset grid method: Here a fluid grid around the solid will move in the fluid. The outer fluid grid will not move and both the grid has to be solved by iteration until convergence. Miller et al. (2014) report that this method is preferable to ALE formulation.

6.2.3 Vortex Shedding

During vortex shedding, the structure oscillates with Strouhal frequency and hence the movement of the structure with respect to thickness of the structure (y/D) ratio is less than 1 as reported before. High amplitude is expected when the structural frequency coincides with Strouhal frequency, as shown in Figure 6.3b. The range of y/D ratio can be estimated from the work of Griffin and Ramberg (1982).

For the VIV study circular cylinders are considered. Wang et al. (2001) used 3D CFD model to study FSI of a beam at Re = 3900. Dong and Karniadakis (2005) investigated an oscillating cylinder at Re = 10 000 using spectral method. The review from these references gives the CFD development in this area.

6.2.4 Galloping of a Rectangular Cylinder

Galloping occurs for cylinders that are not circular. Robertson et al. (2003) and Dettmer and Peric (2006) illustrated the application of CFD with moving grid to compute the critical velocity for galloping of square and rectangular cylinders. Robertson et al. (2003) compared their work for Re = 250 to experimental measurements with Re far greater than 250 and found good matching results.

6.2.5 Bridge Aerodynamics

Flow around bridges is done for two cases. One is to get the pressure coefficient around a fixed bridge and the other for movable bridge. The movable bridge CFD modeling is done to calculate the critical velocity for flutter. Xu (2013) and Larsen and Larose (2015) reviewed dynamic effect of wind on cable-stayed bridges.

6.2.5.1 Fixed Bridge Computation

In the first case, the effect of vortex shedding on pressure coefficient can be calculated for static design. In the second case fluid–structure interaction needs to be considered, and this becomes a far more complicated modeling. In both cases, first challenge is to create proper grid resolution around bridges. Selvam et al. (2002) used finite element methods to locate the flow separation region using 0.001B grid resolution, where B is the width of the bridge. In the recent years, there are sophisticated grid generators and solvers are available using finite difference method (FDM) or finite element method (FEM). Patro et al. (2013) used adaptive FEM to resolve the grid where the numerical error is high. They compared the pressure coefficients of the Great Belt East bridge with WT measurements.

6.2.5.2 Movable Bridge Computation for Critical Flutter Velocity Using Moving Bridge

For computation of critical flutter velocity (U_{cr}), moving grid is used to move the bridge with heaving and pitching motion as discussed in Selvam et al. (2001, 2002). As discussed before, forced motion or free motion method can be used to calculate the flutter velocity as reported in Selvam (2017). Patro et al. (2010) used adaptive finite element method to calculate the flutter velocity of two bridge cross sections using free motion method. To reduce the computer time, Selvam and Bosch (2016) used parallel computation. In the free motion method, when $U < U_{cr}$, the angle of the bridge with respect to horizontal (α) will not increase from the initial perturbation. In the computation reported in Selvam and Bosch (2016), an initial value of $\alpha = 1.8$ degrees is used. For a well-refined grid of 425×90 with 0.001B as the smallest grid close to the Great Belt East Bridge (GBEB) section, the response of the bridge is shown for a reduced velocity U^* ($U/\omega_p B$) of 1.2 and 1.45 are shown in Figure 6.4. Here B is the width, ω_p is the pitching frequency, and U is the approach velocity for the bridge. When $U^* < U_{cr}$(1.4) as in Figure 6.4a, the pitching angle reduces to zero due to positive aerodynamic damping. When U>Ucr, then the angle keep on increasing, as shown in Figure 6.4b. Knowing the value of B, ω_p, and U^*, one can calculate $U_{cr} = \omega_p B U^*$. The computed 74 m/s compares well with WT measurements as reported by Larsen and Walther (1998). When coarse grid is used the U_{cr} reduced using CFD.

The visualization of the GBEB bridge with grid spacing close to the grid is reported in Figure 6.5 for $U^* = 12.5$. The plot at the end of 100 time units for $U^* = 1.45$ is shown in Figure 6.6 for flutter condition.

6.2.5.3 Estimation of Negative Damping Coefficient of a Bridge Considering the Response as a Free Vibration

For free vibration, the damping coefficient ξ can be calculated from the ratio of one cycle amplitude to next cycle from the following equation derived in Paz and Leigh (2004):

$$2\pi\xi = \ln(y_1/y_2)$$

where y_1 and y_2 are consecutive displacement or angle. Using the pitch angle reported in Selvam et al. (2001, 2002) and Selvam and Bosch (2016) during flutter, the damping coefficient is calculated using the preceding equation. The negative damping coefficient ξ ranges from 2 to 9%. Depending upon the incoming velocity speed, shape, and elastic properties of the bridge the amplitude changes.

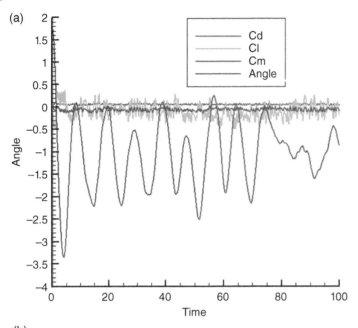

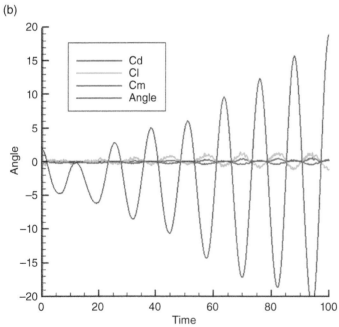

Figure 6.4 Response of the GBEB for (a) $U^* = 1.2$ and (b) $U^* = 1.45$.

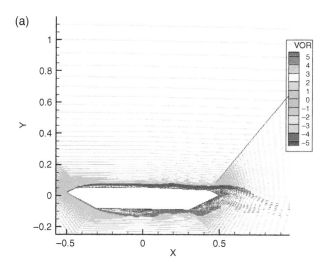

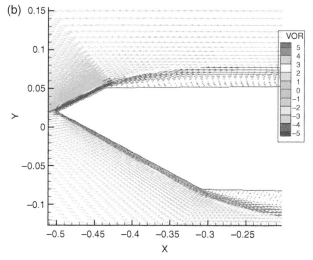

Figure 6.5   Flow around the GBEB bridge for $U^* = 12.5$ (a) full view of the bridge and (b) close up view.

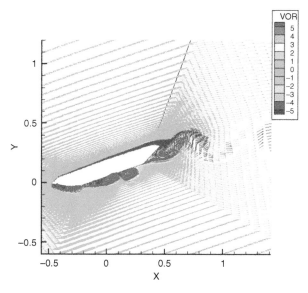

Figure 6.6   Flow around GBEB for $U^* = 1.45$ (flutter condition).

6.3 Inflow Turbulence by Body Forcing

As discussed in the previous chapter, the inflow turbulence is introduced at the inlet using synthetic turbulence generators. To transport all scales, equal grid spacing is necessary. This requires extensive storage and computer time. In addition, when synthetic turbulence is introduced at the inlet, they will not satisfy the continuity equation properly for reasons mentioned in Patruno and de Miranda (2020). That is, even though the time-varying inlet velocity function may satisfy the continuity equation, in the pressure correction step, only normal velocities are specified and other velocities needs to be corrected. Mostly at the inlet, the other velocities will not be corrected due to programming complexities. To circumvent these problems, Gilling et al. (2009) and Sorenson et al. (2015) introduced the turbulence on the upstream plane of the wind turbines as a body force or source term. The turbulence is introduced similar to immersed boundary concept. That is the body force is calculated from the momentum equation coefficients, as shown in Equation (6.1), and the details are explained in Troldborg et al. (2007) and Troldborg (2009). This reduces the amount of grid spacing on the upwind part of the structure. The idea emerged from Spille-Kohoff and Kaltenbach (2001) and further investigated by Keating et al. (2004). New developments were reported by Haywood (2019).

Discretized momentum equation with $\mathbf{U}$ as vector: $Ap\mathbf{Up} + \sum An b\mathbf{Unb} = Sp + \mathbf{f_T}$

Here $\mathbf{f_T}$ is calculated as

$$\mathbf{f_T} = Ap(\mathbf{Uave} + \mathbf{u'}) + \sum An b\mathbf{Unb} - Sp \tag{6.1}$$

6.4 CFD for Improving Wind Turbine Performance and Siting and Wind Tunnel Design

Wind interacting with wind turbines has become an attractive research in the recent years due to need for producing efficient renewable wind energy. Similarly, wind tunnel has been used extensively for many different areas from building and bridge aerodynamics to wind turbine siting in complex terrain. Similarly, from electronic cooling to room cooling, fans are used. In all these areas, a turbine or fan is used either absorbing or producing energy from wind or pump energy into the wind. This can be called as source and sink problem and hence same computational method is applicable for both cases. This section is devoted to discuss some of the methods available.

6.4.1 Actuator Disc Method (ADM)

This method is the easiest one to implement comparing to all the other methods and computationally not intensive. The other methods are more sophisticated and computationally intensive. Accordingly, the physics of the problem considered is more accurate in other methods. In this method, a fan or wind turbine is considered as a circular disc, as shown in Figure 6.7b. In that region, a body force is applied to NS equation.

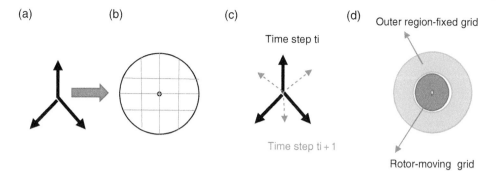

Figure 6.7 Different methods to model a fan or turbine (a) fan or turbine, (b) actuator disc method, (c) actuator line method, and (d) sliding mesh method.

The basics of the method can be explained using Bernoulli equation between two points upwind (point 1) and downwind (point 2) of the actuator disc. Including a source or sink in Bernoulli equation between two points with a pump or turbine in between is

$$p_1 + U_1^2/2 + f_R = p_2 + U_2^2/2$$

In the preceding equation, f_R is positive if energy is pumped in or negative when energy is taken out as in wind turbine. The value of f_R can be determined from performance measurement. A good discussion of the performance curve in a closed-circuit wind tunnel is provided by Moonen et al. (2006). Jimenez et al. (2007, 2008) gives the detail of the equation used for wind turbine. For further details of the momentum theory, one can refer to Sorenson (2016) and Hoem (2018) thesis. Hoem (2018) and Kroll et al. (2015) provided details of OpenFOAM (OF) implementation. The OF case file for Kroll is available in the github website.

6.4.2   Actuator Line Method (ALM)

Here body force is applied along a line, as shown in Figure 6.7c. Here the drag and lift forces created on an airfoil of the rotor due to wind are calculated and then they are applied as body force along a line. Since the blades are moving, the body force along the line is changing with time, as shown in Figure 6.7c. This represents the wind turbine much more accurately as per the discussions in Troldborg et al. (2007) and Sorenson et al. (2015). More details of the implementation are found in the preceding references.

6.4.3   Multiple Reference Frame

This model uses the geometry of the fan blades and therefore does not require any experimental data. The entering velocity is converted to moving reference frame. For details refer to Dogruoz and Shankaran (2017).

6.4.4 Sliding Mesh Model or Rigid Body Motion Model

This method is computationally expensive. Here the rotor region of the grid is moved, and the outer region grid is fixed grid, as shown in Figure 6.7d. Information has to be moved from moving grid to fixed grid. For further references on this topic, one can refer to Ferziger and Peric (2002) and Franzke et al. (2019).

6.4.5 Wind Tunnel Flow Modeling and Design

Moonen et al. (2006) and Calautit et al. (2014) used CFD to understand the flow conditions in a closed loop wind tunnel. They used RANS model for flow computation. Sitek et al. (2017a, b) used multireference frame and sliding mesh method for an open circuit wind tunnel modeling. They used RANS and LES turbulence models.

6.4.6 Improving Wind Turbine Performance

Extensive work is done using CFD to model single and multiple turbines in the computational domain. Further including the terrain effect on wind farm performance is also investigated. Recent review by Sorenson et al. (2017) and Porté-Agel et al. (2020) provides current status in this area. Sorenson et al. (2015) considered inflow turbulence by body force as mentioned earlier.

6.5 Tornado–Structure Interaction

For modeling tornado–structure interaction, first of all one should have proper tornado or vortex model. Then in the tornado model, building can be allowed to interact. First existing tornado model will be discussed and then the status of tornado–structure interaction will be reviewed.

6.5.1 Tornado Models for Engineering Applications

Tornado models used for engineering applications can be classified into two major categories: (i) analytical vortex model and (ii) vortex generation chamber models. Under the vortex generation chamber model, one can have stationary vortex generation chamber and moving vortex generation chamber. In Selvam (2017) review, the vortex generation chamber models are classified separately into two categories.

6.5.2 Analytical Vortex Model

In the analytical tornado vortex model, the tornado at every instant of time is described by a mathematical equation, as presented in Selvam and Millett (2003, 2005). This model considers the tangential velocities and translational velocity of a tornado but not any radial or vertical velocity. The tangential wind profiles prescribed by the Rankine combined vortex and Vatistas model reasonably compares with field and laboratory measurements, and the details are reported in Strasser and Selvam (2015a, b). This model has been used extensively by Selvam and his research group for more than 25 years. They applied the model for various vortex–structure interaction studies as discussed later.

Vortex–2D Cylinder Interaction: To compute time variation of forces and flow features on 2D cylinders as reported in Strasser et al. (2016). In that they also introduced a novel approach to find the period of the tornado vortex for dynamic analysis.

Forces on 3D Cube and Dome: Forces on 3D cube are reported in Selvam and Millett (2003, 2005) and Alrasheedi and Selvam (2011). Selvam and Millett (2005) reported that the *vortex-type interaction on a cubic building produced more than twice the force on the roof comparing to straight wind* for the same reference wind speed. Alrasheedi showed that if the building is larger than the core radius, then the force on the building approaches straight wind.

Yousef et al. (2018) compared the tornado forces on a dome and cube and showed that the roof forces on a cube is 1.8 times more than on a dome. This is merely due to the shape of the structure.

Engineered Wall (Long 2D Wall) as Shelter: Tornado–terrain interaction of 2D hill studies is reported in Gorecki and Selvam (2014, 2015). In this work, they reported that 2D structures can be used as a fence to shelter neighboring building. When the vortex interacts with a 2D fence, the velocity behind the fence is reduced certain distance, and this provides shelter against tornado type wind flow. Further details can be found in the preceding references.

Terrain Effect on the Tornado Path: Ahmed and Selvam (2015) investigated the tornado path deviation when a vortex passes over a ridge. They used 3D finite element model to model the flow. They compared their results with experimental measurements of Karstens (2012) and field observations of different tornado using Google Earth.

In the field observation of Mayflower tornado when a tornado passes a hill at an angle, Selvam et al. (2015a, b) reported that at certain instances the tornado went on one side of a hill (path 1) and certain other instances the tornado went on the other side of the hill (path 2), as shown in Figure 6.8a. To find the reason for this to happen, Dominquez and Selvam (2016, 2017) used 2D vortex model to investigate the vortex–hill or rectangular cylinder investigation. In this study, they released the vortex at different angles and perpendicular distance (TDIST) to the hill and observed the interaction. They found that the path of the tornado angle and TDIST determines which way the tornado will pass the hill. Dominquez and Selvam (2016) considered range of parameters to suit the Mayflower field observation.

Here a hill size of $W \times 6W$, where W is the width of the hill, a tornado size of $r_{max} = 2D$, and Rankine vortex strength α as 1 are considered for interaction here. This gives a maximum velocity of $U_T + \alpha\, r_{max}$ where U_T is the translational velocity of the tornado. Using these parameters, the vortex is released at an angle of 15 degrees with TDIST $-D$, $-1.5D$, $-2D$, and $-2.5D$ with respect to the center of the hill. Only for $-2.5D$ case, the vortex went below the hill and for all the other cases the vortex went above the hill. For illustration purposes, TDIST equals to $-1.5D$, and $-2.4D$ cases are considered in Figure 6.9.

From Figure 6.9, one can see that when the vortex is traveling below the hill (path 1), there is vortex on both sides of the hill, and when the vortex is traveling above the hill, there is vortex on one side, i.e., on the top of the hill. So, for path 1, there is damage on both sides of the hill but not for path 2. Similar damages were noticed from the Google Earth

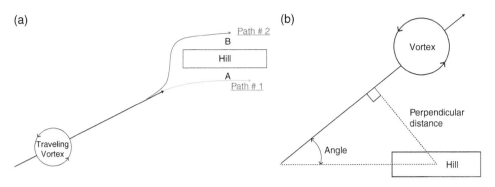

(a)

Path # 2

B

Hill

A

Path # 1

Traveling
Vortex

(b)

Vortex

Perpendicular
distance

Angle

Hill

Figure 6.8 Tornado paths with respect to a hill (a) different paths and (b) influencing parameters: angle of the path and perpendicular distance from the hill.

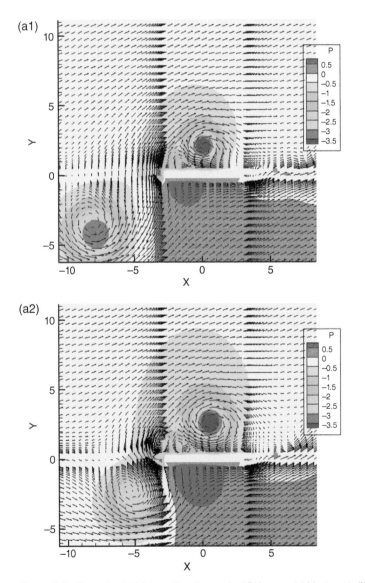

Figure 6.9 Tornado–hill interaction for angle 15 degree. (a) Vortex at different times for path 1 or vortex traveling below the hill for TDIST = −2.5D. (b) Vortex at different times for path 2 or vortex traveling above the hill for TDIST = −1.5D

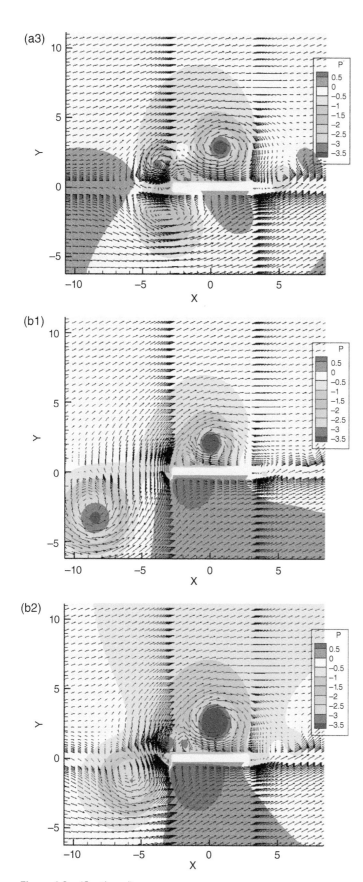

Figure 6.9 (Continued)

(b3)

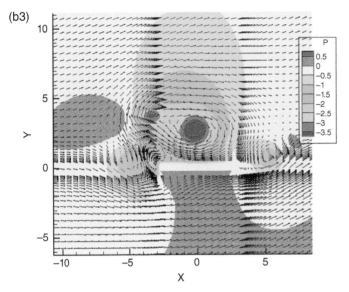

Figure 6.9 (Continued)

observation of Mayflower damage site. For the path 1, the vortex strength below the hill is less compared to the vortex strength above the hill in part 2.

6.5.3 Vortex Generation Chamber Models

In the tornado vortex generation chamber or tornado chamber (TC) models, a tornado-like vortex is developed by providing tangential velocity V_θ and radial velocity V_r around a circular cylinder as inlet velocity and vertical velocity at the top of the circular hole as outlet velocity, as shown in Figure 6.10a and b. In a stationary vortex generation chamber, the building is moved and the chamber is kept stationary, as shown in Figure 6.10a; and in the moving vortex chamber, the chamber is moved and building is kept stationary, as shown

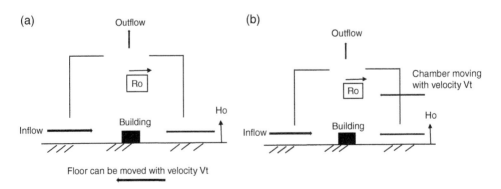

Figure 6.10 (a) Stationary vortex chamber. (b) Moving vortex chamber.

in Figure 6.10b. By controlling the V_θ, V_r, top radius Ro and the spacing between the floor and vortex chamber Ho, and several types of tornado vortex can be developed in the vortex chamber. The preceding factors are related by the term swirl ratio and is defined as

$$S = V_\theta/(2AV_r) = \tan\theta/2A$$

where aspect ratio A=Ho/Ro and $\tan\theta = V_\theta/V_r$.

6.5.3.1 Stationary Vortex Chamber

Computer modeling work is conducted by Lewellen and Lewellen (1997), Hangan and Kim (2008), and Natarajan and Hangan (2012) to study the wind field due to the effect of different parameters. Recently, Lewellen (2010) investigated the tornadic flow over hill using immersed boundary method. Here they moved the hill with respect to the simulator. The major drawback with this approach is that there is no proper grid resolution on the surface of the hill and hence the investigation of the tornado–hill interaction may not be that accurate. Zu et al. (2016) studied the flow and pressure around a dome due to tornado by moving the dome. Herethey moved the dome by dynamic mesh method. They did not use enough grid resolution close to the building and also no comparison with tunnel study. Nasir et al. (2014) studied the flow around a building by placing the building at three different locations in the TC. They did not report any detailed comparison of the force and pressure coefficients with straight wind flow.

   Kashefizadeh et al. (2019) used an axisymmetric model to show that when the width of the tornado chamber increases, the touchdown swirl ratio (S_{TD}) and location of the maximum tangential velocity increases. This study is only parametric study without one to one comparison to field or experimental measurements. Verma and Selvam (2020) introduced an efficient 3D model to compute and validate the flow in the Texas Tech University (TTU) tornado chamber, as shown in Figure 6.11. This model eliminates the complex modeling of inside and outside of the tornado chamber, as shown in Figure 6.11a to a simplified modeling of only in the inner chamber, as shown in Figure 6.11b. The details of the modeling and boundary conditions are reported in Verma and Selvam (2020). Because of this, they could have refined grid in the core region and also this eliminates the modeling of vane and other parts at the inlet of the chamber. The latter approach is used by Fangping et al. (2016) and Gairola and Bitsuamlak (2019). The computed STD value of 0.22 by Verma et al. is in very good comparison with (TC) measurements. They also compared the V_{max}, z_{max}, and r_{max} with TC measurements, and they are also in reasonable comparison. They recommended these four basic parameters to be validated with TC for further applications.

6.5.3.2 Moving Vortex Chamber

Kuai et al. (2008), Sengupta et al. (2008), and Phuc et al. (2012) moved the vortex chamber utilizing sliding mesh algorithm. Kuai et al. (2008) reported only wind field study and Sengupta et al. (2008) reported CFD study for a microburst interacting on a cube. Only Phuc et al. in (2012) reported computed maximum and minimum pressure coefficients on a cube for a swirl ratio of 0.65. They did not have any comparison with TC study.

(a)

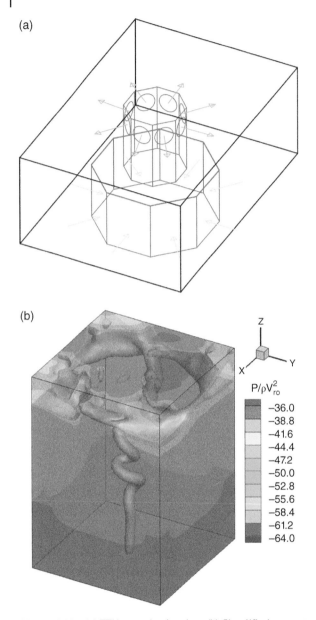

(b)

$P/\rho V_{ro}^2$

- −36.0
- −38.8
- −41.6
- −44.4
- −47.2
- −50.0
- −52.8
- −55.6
- −58.4
- −61.2
- −64.0

Figure 6.11 (a) TTU tornado chamber. (b) Simplified geometry for CFD model of TTU tornado chamber.

6.6 Wind Environment Around Buildings

Mostly, RANS equations are used. Especially k–ε turbulence model is used extensively. To calculate the mean velocity, this method is sufficient. Also, this method takes less computer time (may be an order less comparing to LES) when the steady state NS equation is solved by

SIMPLE-type solvers. There are many papers validating the CFD model in the literature. As an example, Liu et al. (2018) used k-ε turbulence model to validate the CFD results with field measurement for a computational region of 12.6 km × 5.4 km × 0.351 km. They used simplified and detailed models and compared the results with measurements. They addressed the roughness length effect on the CFD results.

6.7 Pollutant Transport Around Buildings

First flow around building needs to done and then pollutant transport calculation is done. This could be steady or unsteady flow computation. Recent reviews like Lateb et al. (2016) and Tominaga and Stathopoulos (2016) can be looked in for further details.

6.8 Parallel Computing for Wind Engineering

CFD modeling using LES takes lots of computer time. To reduce the clock time or turna-round time, parallel computing is very useful. Selvam and his group (Ahmed and Selvam 2015; Ahmed 2016, and Yousef et al. 2018) used parallel computing for investigating the vortex–hill and vortex–dome interactions. The opensource program OpenFOAM can use the parallel computing once the case file is set for serial computing. Other commercial programs also come under the same category.

6.9 Chapter Outcome

Preliminary attempt is made to expose CFD areas that are not covered from Chapters 1–5 and are relevant to wind engineering. There may be other topics that may be relevant and individual researcher's needs to do their own reviews.

References

Ahmed, N.S. (2016). Field observations and computer modeling of tornado-terrain interaction and its effects on tornado damage and path. PhD dissertation, University of Arkansas.

Ahmed, N.S. and Selvam, R.P. (2015). Ridge effects on tornado path deviation. *International Journal of Civil and Structural Engineering Research* 3 (1): 273–294.

Alrasheedi, N.H. and Selvam, R.P. (2011). Tornado forces on different building sizes using computer modeling. *ASME Early Career Technical Conference*, Fayetteville, AR, USA (31 March–2 April 2011).

Blocken, B. (2014). 50 years of computational wind engineering: Past, present and future. *Journal of Wind Engineering & Industrial Aerodynamics* 129: 69–102.

Blocken, B. (2018). LES over RANS in building simulation for outdoor and indoor applications: a foregone conclusion? *Building Simulation* 11: 821–870.

Calautit, J.K., Chaudhry, H.N., Hughes, B.R., and Sim, L.F. (2014). A validated design methodology for a closed-loop subsonic wind tunnel. *Journal of Wind Engineering and Industrial Aerodynamics* 125: 180–194.

Cao, H.L., Quan, Y., Gu, M., and Wu, D. (2012). Along-wind aero dynamic damping of isolated rectangular high-rise buildings. *Journal of Vibration and.Shock* 31: 122–127.

Cheng, C.M., Lu, P.C., and Tsai, M.S. (2002). Across wind aerodynamic damping of isolated square-shaped buildings. *Journal of Wind Engineering and Industrial Aerodynamics* 90: 1743–1756.

Davenport, A.G. (1995). How can we simplify and generalize wind loads? *Journal of Wind Engineering and Industrial Aerodynamics* 54 (55): 657–669.

Davenport, A.G. and Novak, M. (1976). *Vibration of Structures Induced by Wind, from Harris' Shock and Vibration Hand Book*. New York: McGraw Hill Chapter 29.

Dettmer, W.G. and Perić, D. (2006). A computational frame work for fluid rigid body interaction: finite element formulation and applications. *Computer Methods in Applied Mechanics and Engineering* 195: 1633–1666.

Dogruoz, M.B. and Shankaran, G. (2017). Computations with the multiple reference frame technique: Flow and temperature fields downstream of an axial fan. *Numerical Heat Transfer Part A Applications* 71: 488–510.

Dominguez, D. and Selvam, R.P. (2017). Tornado width for computer modeling from Google Earth data and period of the vortex, *Proceedings of the ASCE Architectural Engineering Institute's 2017 Conference (AEI 2017)*, Oklahoma City, OK (12–14 April 2017), 470–483

Dominquez, D., and R.P. Selvam (2016), Two-dimensional simulation of tornado-hill interaction using CFD, from *Advances in Mechanics and Materials*, Eds. A.N. Nayak, S.K. Patro and R. Panigrahi. International Publishing House, New Delhi, p. 107–112. *Proceedings: International Conference on Recent Advances in Mechanics and Materials (ICRAMM-2016)* (17–18 December 2016), Veer Surendra Sai University of Technology (VSSUT), Burla, Odisha, India, 107-112.

Dong, S. and Karniadakis, G.E. (2005). DNS of flow past a stationary and oscillating cylinder at Re=10000. *Journal of Fluids and Structures* 20: 519–531.

Fangping, Y., Yan, G., Honerkamp, R. et al. (2016). Effects of chamber shape on simulation of tornado-like flow in a laboratory. In: *Wind Engineering for Natural Hazards Modeling, Simulation, and Mitigation of Windstorm Impact on Critical Infrastructure* (ed. A. Mousaad Aly and E. Dragomirescu). Reston, VA: ASCE.

Ferziger, J.H. and Peric, M. (2002). *Computational methods for fluid dynamics*, 3ee. Springer.

Franzke, R., Sebben, S., Bark, T. et al. (2019). Evaluation of the multiple reference frame approach for the modelling of an axial cooling fan. *Energies* 12: art. no 2934.

Gairola, A. and Bitsuamlak, G. (2019). Numerical tornado modeling for common interpretation of experimental simulators. *Journal of Wind Engineering and. Industrial Aerodynamics* 186 (March): 32–48.

Gilling, L., Sørensen, N., and Rethore, P. (2009). Imposing resolved turbulence by an actuator in a detached eddy simulation of an airfoil. In: *Proceedings. of EWEC 2009, Marseille, France, 16–19 March 2009*. Brussels, Belgium: European Wind Energy Association.

Gorecki, P.M. and Selvam, R.P. (2014). Visualization of tornado-like vortex interacting with wide tornado-break wall. *Journal of Visualization* 18: 393–406.

Gorecki, P.M. and Selvam, R.P. (2015). Rankin combined vortex interaction with rectangular prism. *International Journal of Computational Fluid Dynamics* 29: 120–132.

Griffin, O.M. and Ramberg, S.E. (1982). Some recent studies of vortex shedding with application to marine tubulars and risers. *ASME. Journal of Energy Resources Technology* 104 (1): 2–13.

Hangan, H. and Kim, J.D. (2008). Swirl ratio effects on tornado vortices in relation to the Fujita scale. *Wind and Structures* 11: 291–302.

Haywood, J.S. (2019). Turbulent inflow generation methods for Large Eddy Simulations. PhD thesis, Mississippi State University.

Hoem, M.E. (2018), Wind turbine simulations with OpenFOAM, Master thesis, Norwegian University of Science and Technology.

Ho-Le, K. (1988). Finite element mesh generation methods: a review and classification. *Computer-Aided Design* 20: 27–38.

Jimenez, A., Crespo, A., Migoya, E., and Garcia, J. (2007). Advances in large-eddy simulation of a wind turbine wake. *Journal of Physics: Conference Series* 75: 012041.

Jimenez, A., Crespo, A., Migoya, E., and Garcia, J. (2008). Large-eddy simulation of spectral coherence in a wind turbine wake. *Environmental Research Letters* 3: 015004.

Karstens, C.D. (2012). Observations and laboratory simulations of tornadoes in complex topographical regions. PhD dissertation, Department of Meteorology, Iowa State University.

Kashefizadeh, M.H., Verma, S., and Selvam, R.P. (2019). Computer modelling of close-to-ground tornado wind-fields for different tornado widths. *Journal of Wind Engineering and Industrial Aerodynamics* 191: 32–40.

Keating, A., Piomelli, U., Balaras, E., and Kaltenbach, H.-J. (2004). A priori and a posteriori tests of inflow conditions for large-eddy simulation. *Physics of Fluids* 6: 4696–4712.

Kroll, T., Hall, W., and Gagnon, I. (2015). Offshore turbine arrays: numerical modeling and experimental validation. *Honors Theses and Capstones* 224: https://scholars.unh.edu/honors/224.

Kuai, L., Haan, F.L., Gallus, W.A., and Sarkar, P.P. (2008). CFD simulations of the flow field of a laboratory simulated tornado for parameter sensitivity studies and comparison with field measurements. *Wind and Structures* 11: 75–96.

Larsen, A. and Larose, G.L. (2015). Dynamic wind effects on suspension and cable-stayed bridges. *Journal of Sound and Vibration* 334: 2–28.

Larsen, A. and Walther, J.H. (1998). Discrete vortex simulation of flow around five generic bridge deck sections. *Journal of Wind Engineering and. Industrial. Aerodynamics* 77–78: 591–602.

Lateb, M., Meroney, R.N., Yataghene, M. et al. (2016). On the use of numerical modelling for near-field pollutant dispersion in urban environments – a review. *Environmental Pollution Part A* 208: 271–283.

Le Maître, O.P., Scanlan, R.H., and Knio, O.M. (2003). Estimation of the flutter derivatives of an NACA airfoil by means of Navier–Stokes simulation. *Journal of Fluids and Structures* 17: 1–28.

Lewellen, D. C., (2010). Effects of topography on tornado dynamics: a simulation study. Preprints, *26th Conference on Severe Local Storms*, AMS, Nashville, TN, paper 4B.1.

Lewellen, W.S. and Lewellen, D.C. (1997). Large-Eddy simulation of a tornado's interaction with the surface. *Journal of the Atmospheric Sciences* 54: 581–605.

Liu, S., Pan, W., Zhao, X. et al. (2018). Influence of surrounding buildings on wind flow around a building predicted by CFD simulations. *Building and Environment* 140: 1–10.

Lo, D.S.H. (2015). *Finite Element Mesh Generation*. New York: CRC Press.

Miller, S.T., Campbell, R.L., Elsworth, C.W. et al. (2014). An Overset grid method for fluid-structure interaction. *World Journal of Mechanics* 4: 217–237.

Mittal, S. and Tezduyar, T.E. (1992). A finite element study of unsteady incompressible flows past oscillating cylinders and airfoils. *International Journal for Numerical Methods in Fluids* 15: 1073–1118.

Moonen, P., Blocken, B., Roels, S., and Carmeliet, J. (2006). Numerical modeling of the flow conditions in a closed-circuit low speed wind tunnel. *Journal of Wind Engineering and Industrial Aerodynamics* 94: 699–723.

Nasir, Z., Bitsuamlak, G.T., and Hangan, H. (2014). Computational modeling of tornadic load on a building. *6th International Symposium on Computational Wind Engineering*, Hamburg, Germany (8–12 June).

Natarajan, D. and Hangan, H. (2012). Large eddy simulations of translation and surface roughness effects on tornado-like vortices. *Journal of Wind Engineering and Industrial Aerodynamics* 104–106: 577–584.

Parkinson, G. (1989). Phenomena and modelling of flow-induced vibrations of bluff bodies. *Progress in Aerospace Sciences* 26: 169–224.

Patro, S.K., Selvam, R.P., and Bosch, H.R. (2010). Bridge flutter modeling using H–adaptive FEM. *Journal of Wind and Engineering* 7: 39–48.

Patro, S.K., Selvam, R.P., and Bosch, H. (2013). Adaptive h-finite element modeling of wind flow around bridges. *Engineering Structures* 48: 569–577.

Patruno, L. and de Miranda, S. (2020). Unsteady inflow conditions: a variationally based solution to the insurgence of pressure fluctuations. *Computer Methods in Applied Mechanics and Engineering* 363: 112894.

Paz, M. and Leigh, W. (2004). *Structural dynamics: Theory and Computation*, 5ee. Kluwer Academic Publishers.

Phuc, P.V., Nozu, T., Nozawa, K. and Kikuchi, H. (2012). A numerical study of the effects of moving tornado-like vortex on a cube. *The Seventh International Colloquium on Bluff Body Aerodynamics and Applications (BBAA7)*, Shanghai, China (2–6 September).

Porté-Agel, F., Bastankhah, M., and Shamsoddin, S. (2020). Wind-turbine and wind-farm flows: a review. *Boundary-Layer Meteorol* 174: 1–59.

Robertson, I., Li, L., Sherwin, S.J., and Bearman, P.W. (2003). A numerical study of rotational and transverse galloping rectangular bodies. *Journal of Fluids and Structures* 17: 681–699.

de Sampaio, P.A.B., Lyra, P.R.M., Morgan, K., and Weatherill, N.P. (1993). Petrov-Galerkin solutions of the incompressible Navier-Stokes equations in primitive variables with adaptive remeshing. *Computer Methods in Applied Mechanics and Engineering* 106: 143–178.

Selvam, R.P. (2008). Developments in computational wind engineering. *Journal of Wind and Engineering* 5: 47–54.

Selvam, R.P., (2010). Building and bridge aerodynamics using computational wind engineering. *Proceedings: International Workshop on Wind Engineering Research and Practice*, Chapel Hill, NC, USA (28–29 May).

Selvam, R.P. (2017). CFD as a tool for assessing wind loading. *The Bridge and Structural Engineer* 47 (4): 1–8.

Selvam, R.P. and Bosch, H. (2016). Effect of initial conditions and grid refinements on bridge flutter calculations using CFD and HPC. *Proceeding: 8th International Colloquium on Bluff Body Aerodynamics and Applications*, Northeastern University, Boston, Massachusetts, USA (7–11 June).

Selvam, R.P. and Millett, P.C. (2003). Computer modeling of tornado forces on buildings. *Wind & Structures* 6: 209–220.

Selvam, R.P. and Millett, P.C. (2005). Large eddy simulation of the tornado-structure interaction to determine structural loadings. *Wind & Structures* 8: 49–60.

Selvam, R.P., Govindaswamy, S. and Bosch, H. (2001), Aeroelastic analysis of bridge girder section using computer modeling. Final report: MBTC FR-1095, Mack-Blackwell National Rural Transportation Study Center, University of Arkansas. http://ntl.bts.gov/lib/11000/11100/11186/1095.pdf.

Selvam, R.P., Govindaswamy, S., and Bosch, H. (2002). Aeroelastic analysis of bridges using FEM and moving grids. *Wind & Structures* 5: 257–266.

Selvam, R.P., Lin, L., and Ponnappan, R. (2006). Direct simulation of spray cooling: effect of vapor bubble growth and liquid droplet impact on heat transfer. *International Journal of Heat and Mass Transfer* 49: 4265–4278.

Selvam, R.P., Ahmed, N., Strasser, M.N., et al. (2015a). RAPID: documentation of tornado track of Mayflower Tornado in hilly terrain. NSF Final Report. Department of Civil Engineering, University of Arkansas.

Selvam, R.P., Strasser, M.N., Ahmed, N. et al. (2015b). Observations of the influence of Hilly Terrain on tornado path and intensity from the damage investigation of the 2014 Tornado in Mayflower, Arkansas. *Proceedings: Structures Congress* 2015: 2711–2721.

Sengupta, A., Haan, F.L., Sarkar, P.P., and Balaramudu, V. (2008). Transient loads on buildings in microbusts and tornado winds. *Journal of Wind Engineering and Industrial Aerodynamics* 96: 2173–2187.

Simiu, E. and Scanlan, R.H. (1986). *Wind Effects on Structures: An Introduction to Wind Engineering*. New York: Wiley.

Sitek, M., Bojanowski, C., Lottes, S. A., and Bosch, H. (2017a), Validation of CFD Models for a Virtual Aerodynamics Laboratory. *The 13th Americas Conference on Wind Engineering (13ACWE)*, Gainesville, Florida USA (21–24 May).

Sitek, M. A., Lottes, S. A., and Bojanowski, C. (2017b). Air flow modeling in the wind tunnel of the FHWA Aerodynamics Laboratory at Turner-Fairbank Highway Research Center. https://www.osti.gov/servlets/purl/1401969.

Sørensen, J.N., Mikkelsen, R.F., Henningson, D.S. et al. (2015). Simulation of wind turbine wakes using the actuator line technique. *Philosphical Transactions of the Royal Society A* 373: 20140071.

Sørensen, J.N., Hansen, K.S., Sarmast, S., and Ivanell, S. (2017). A survey of modelling methods for high-fidelity wind farm simulations using large eddy simulation. *Philosphical Transactions of the Royal Society A* 375: 20160097.

Sorenson, J.N. (2016). *General Momentum Theory for Horizontal Axis Wind Turbines*. Springer.

Spille-Kohoff, A. and Kaltenbach, H. J. (2001). Generation of turbulent inflow data with a prescribed shear-stress profile. *Third AFOSR International Conference on DNS/LES Arlington*, TX (5–9 August 2001), in DNS/LES Progress and Challenges (Liu, C., Sakell, L., and Beutner, T., eds).

Strasser, M.N. and Selvam, R.P. (2015a). A Review of viscous vortex tangential velocity profiles for application in CFD. *Journal of the Arkansas Academy of Sciences* 69: 88–97.

Strasser, M.N. and Selvam, R.P. (2015b). The variation in the maximum loading of a circular cylinder impacted by a 2D vortex with time of impact. *Journal of Fluids and Structures* 58: 66–78.

Strasser, M.N., Yousef, M.A.A., and Selvam, R.P. (2016). Defining the vortex loading period and application to assess dynamic amplification of tornado-like wind loading. *Journal of Fluids and Structures* 63: 188–209.

Thompson, J.F., Warsi, Z.U.A., and Mastin, C.W. (1985). *Numerical Grid Generation: Foundations and Applications*. North Holland.

Tominaga, Y. and Stathopoulos, T. (2016). Ten questions concerning modeling of near-field pollutant dispersion in the built environment. *Building and Environment* 105: 390–402.

Troldborg, N. (2009), Actuator Line Modeling of Wind Turbine Wakes. PhD thesis, Technical University of Denmark.

Troldborg, N., Sørensen, J.N., and Mikkelsen, R. (2007). Actuator line simulation of wake of wind turbine operating in turbulent inflow. *Journal of Physics: Conference Series* 75: 012063.

Verma, S. and Selvam, R.P. (2020). CFD to VorTECH pressure-field comparison and roughness effect on flow. *Journal of Structural Engineering* 146 (9): 04020187-1 to 12.

Walther, J.H. (1994). Discrete vortex method for 2D flow past bodies of arbitrary shape undergoing prescribed rotary and translational motion. PhD thesis, Department of fluid mechanics, Technical University of Denmark, Denmark.

Wang, X.Q., So, R.M.C., and Liu, Y. (2001). Flow-induced vibration of an Euler-Bernoulli beam. *Journal of Sound and Vibration* 243: 241–268.

Xu, Y.L. (2013). *Wind Effects on Cable-Supported Bridges*. John-Wiley.

Yousef, M.A.A., Selvam, R.P., and Prakash, J. (2018). A comparison of the forces on dome and prism for straight and tornadic wind using CFD model. *Wind & Structures* 26: 369–382.

Zhang, Y., Habashi, W.G., and Khurram, R.A. (2015). Predicting wind-induced vibrations of high-rise buildings using unsteady CFD and modal analysis. *Journal of Wind Engineering and Industrial Aerodynamics* 136: 165–179.

Zu, J., Yan, G., and Li, C. (2016). Investigation of wind pressure of translating tornado on spherical dome structures. *Proceeding: 8th International Colloquium on Bluff Body Aerodynamics and Applications*, North eastern University, Boston, Massachusetts, USA (7–11 June).

7

Introduction to OpenFOAM Application for Wind Engineering (an Open-Source Program)

R. Panneer Selvam, Sumit Verma, and Zahra Mansouri

7.1 Introduction to OpenFOAM and ParaView for Wind Engineering

7.1.1 OpenFOAM for Wind Engineering

There are several commercial programs available in the market for computational fluid dynamics (CFD) computation and for grid generation. They are very costly and one may have to pay more than $10k/year for one seat. There are also several open-source CFD programs and grid generators available in the web. The listing of several commercial and open-source CFD programs and grid generators and their capabilities are discussed by Liu and Zhang (2019). The popular open-source program used in wind engineering is OpenFOAM. This program is constantly upgraded and new features are introduced by the users. In this chapter, the installation and application of OpenFOAM for flow over building is illustrated for hands-on learning. The book by Moukalled et al. (2016) discusses some of the features of OpenFOAM, but the hands-on part is not that clear. The OpenFOAM basic training notes (OFBT2019) has 14 tutorials, and it is a very good resource available from web.

7.1.2 Grid Generation

As discussed in the previous chapter, there are several methods to generate grid and many commercial and open-source programs are available. The listing of the available programs commercially or otherwise is discussed in Liu and Zhang (2019). Here, we will use the Open-FOAM grid generator. The different grid generators available from OpenFOAM are block-Mesh, snappyHexMesh, and foamyHexMesh. For illustration, we will use blockMesh and snappyHexMesh. The last one is left for individual learning. Verma (2020) report explains how to generate the snappyHexMesh around a building using OpenFOAM. There is extensive information on each method available in the web and also at the OpenFOAM web page.

7.1.3 Visualization

Visualization is another important step as discussed before. For the class teaching at the University of Arkansas, we used Tecplot. This is a commercial program and it is very user-friendly. The open-source program ParaView is recommended for OpenFOAM.

Computational Fluid Dynamics for Wind Engineering, First Edition. R. Panneer Selvam.
© 2022 John Wiley & Sons Ltd. Published 2022 by John Wiley & Sons Ltd.

7.2 Installation of OpenFOAM, ParaView, and Running a Sample File

7.2.1 Installation of OpenFOAM and ParaView

The OpenFOAM runs under Linux operating system, but nowadays this can also be installed in the Windows operating system. The command environment in the OpenFOAM is Linux and when it is run in Windows, the Linux commands are emulated. For long runs, it is recommended to use the Linux system and run them in the background. If the Linux operating system is put together only for computing, then the cost of the system will be cheaper. Since most of them have access to Windows, this chapter is discussing how to install in Windows operating system. Extensive information is available in the web for different systems.

As we discussed in the previous chapters, visualization of the CFD runs is very important part of CFD applications. There are several commercial and open-source visualization programs are available. For detail of the available commercial and open-source programs, one can read about it in Liu and Zhang (2019). For generated data from OpenFOAM, one can visualize it in the OpenFOAM environment using paraFOAM, but the preferred approach is to download ParaView separately in the Windows and visualize the data.

The **preferred installation approach** is to install in Linux or Windows and perform the CFD calculations using OpenFOAM and then visualize using ParaView in the Windows system. For this ParaView, software has to be installed independently. The graphics is much better in the Windows system and found very friendly.

Here are the steps to install:

1) View the video for Windows 10 installation: https://www.youtube.com/watch?v=xj0dB_PsElg.
2) Install Bash on Ubuntu on Windows and run OpenFOAM under Linux system. Create an account with a password.
3) Install OpenFOAM from the following web page by copying the recent tar file into Windows: https://www.openfoam.com/download/openfoam-installation-on-windows-10 (I installed version 2012. Here 2012 means year 2020 and month 12).
4) Locate and Expand in Linux environment.
5) Run a sample problem under Linux environment.
6) Install ParaView in Windows from: https://www.paraview.org/download/.
7) Visualize the OpenFOAM files in Windows.

A more detailed illustration of how to install OpenFOAM in Windows is available from Verma et al. (2021) and in the report ASEE-Introduction.pdf provided at ASEE21-Open-FOAM-Introduction/ASEE-introduction.pdf. In the earlier-mentioned report, flow around a cubic building is considered for illustration, and the case file is available at GitHub – rpsuark/ASEE21-OpenFOAM-Introduction. To learn more about OpenFOAM, one can also download the user guide and other material from Overview (openfoam.com).

7.2.2 Running a Problem Using OpenFOAM

In the OpenFOAM, a problem is run by creating a case file. Let us call the case file we illustrate here as **building**. In this flow around building, input data is created. In the case file **building**, we have **constant**, **system,** and **0** directories. The functions of each of the directories and

Figure 7.1   Details of **building** data or
case file.

building
>   **constant**
>   >   transportProperties
>   **system**
>   >   blockMeshDict
>   >   controlDict
>   >   fvSchemes
>   >   fvSolution
>   **0** [initial and boundary values for velocities and p]
>   >   p
>   >   U

the detail of the files under them are explained in Figure 7.1. The directories are shown in bold letters and files are shown with regular font. Comments are provided within square brackets. The details of the files are listed in the coming sections with explanations.

First, the mesh or grid is created by typing the following command under the directory **building**:

$blockMesh

Once the grid is created, the grid can be viewed using ParaView and the details are discussed later. When the blockMesh command is run, other files are created and the details will be discussed later. Next, the CFD solver is run using the following command under the same directory **building**:

$icoFoam >screen.txt &

Here, "icoFoam" is the particular command to identify particular NS solver used, "screen.txt" file stores the screen output, and "&" sign means here it is run in the background. This way one can do other things in the Linux screen. When "blockMesh" command is used, we did not run in the background because it does not take much computer time, but the CFD computation can take several hours or days depending on the problem. Since we are using Windows, we do not need to use the "nohup" command, but it is used in Linux systems if one logs out. The icoFoam NS solver is used for unsteady solution of continuity and momentum equations. The OpenFOAM says that it is for laminar flow. The best way to say it is direct simulation.

Several case files are kept in OpenFOAM under the tutorial directory. One can learn a lot from those case files. If someone is interested in the application of CFD, they can read the Section 7.3.2 and then move to Section 7.6 for visualization of the output data using ParaView.

7.3   CFD Solvers and Explanation of Input File for Flow Around a Cube

7.3.1   Numerical Schemes and Solvers for the NS equation

The finite volume method is used to approximate the NS equation. The program is applicable for structured and unstructured grid system. One can choose numerical method like central differencing or upwinding to approximate the convection equation depending on the application. Here, central difference is used for our application. The NS equation is solved either by SIMPLE or PISO algorithm. The SIMPLE is used for steady and unsteady state solution and PISO is used for unsteady state solution. The formed simultaneous equations are solved by preconditioned conjugate gradient (PCG) or geometric algebraic multigrid (GAMG) methods. For further detail of this section, one can refer to Chapter 6 of the OpenFOAM user manual.

Several solvers like simpleFoam, icoFoam, etc., are available from OpenFOAM. In this application, icoFoam is used. This is for unsteady state solution. The simpleFoam is for steady state solution. For unsteady state solution, there are several solvers listed in the user manual. The type of solver selected for a particular problem goes with type of turbulene method used.

7.3.2 Flow Around a Cube Using Uniform Inflow

As a first example, flow around a cubic building with a uniform inflow of U = 1 is considered. This is not the proper inflow, but it is easier to implement for illustration. Also, no turbulence model is considered and the Reynolds number Re is kept as 100. The computational region is 12H × 7H × 5H where H is the height of the cube as shown in Figure 7.2. The x direction is kept along the uniform flow and y direction is perpendicular to the x direction in the plan view. The z-direction is kept in the vertical direction from the bottom wall.

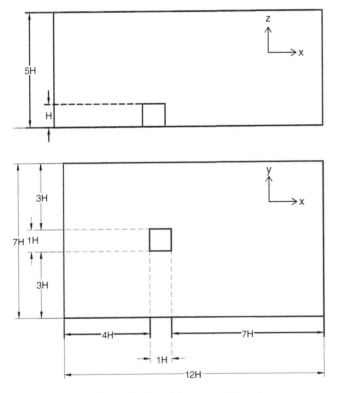

Figure 7.2 Elevation and plan of the computational region.

(a)

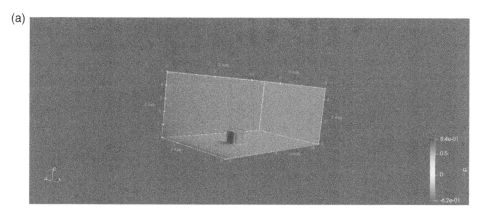

(b)

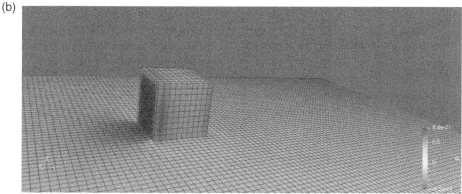

Figure 7.3 (a) 3D view of the computational domain with axes and (b) grid spacing on the building and bottom wall.

The distances of the inlet, outlet, and top boundary from the cube are shown in terms of building height H. Here, H is considered to be one. The Re is calculated as $UH/\nu = 100$. The origin is kept at the left end bottom corner of the computational domain. A 3D view with axes is shown in Figure 7.3 using ParaView as well as the grid features on the building and bottom wall.

Let us look at the detail of the files in each directory of the **building** case file in the following sections.

7.3.3 Detail of "constant" Directory

The **constant** directory has only one file "transportProperties" and is shown in Figure 7.4. The kinematic viscosity ν is given as nu = 1/Re = 0.01. The seven numbers in the square brackets determine the dimension of nu. For more details, refer to user manual Section 2.2.6. After the "blockMesh" command is used to get the grid as discussed before, **polyMesh** directory is created. This will have five different files (boundary, faces, neighbor, owner, and points). After creating the mesh, one can visualize using ParaView.

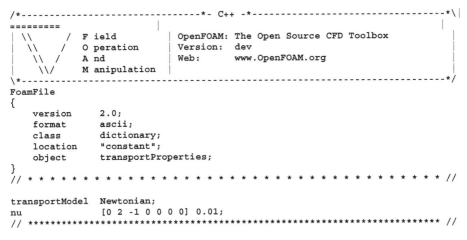

```
/*--------------------------------*- C++ -*----------------------------------*\|
| =========                 |                                                  |
| \\      /  F ield         | OpenFOAM: The Open Source CFD Toolbox            |
|  \\    /   O peration     | Version:  dev                                    |
|   \\  /    A nd           | Web:      www.OpenFOAM.org                       |
|    \\/     M anipulation  |                                                  |
\*---------------------------------------------------------------------------*/
FoamFile
{
    version     2.0;
    format      ascii;
    class       dictionary;
    location    "constant";
    object      transportProperties;
}
// * * * * * * * * * * * * * * * * * * * * * * * * * * * * * * * * * * * * * //

transportModel  Newtonian;
nu              [0 2 -1 0 0 0 0] 0.01;
// *************************************************************************** //
```

Figure 7.4 Detail of file "transportProperties" under **constant** directory.

7.3.4 Detail of "0" Directory

The initial and boundary conditions for pressure and velocities are provided in the files "p" and "U" under **0** directory. The velocity file is listed in Figure 7.5 and pressure file is listed in Figure 7.6. Only at the inlet, a velocity of 1 is provided and the other values are kept 0. Inside the domain, the values of the velocities are kept 0. Similarly, pressure value is kept 0 as a start. The boundary conditions at each face and wall are specified in these files.

7.3.5 Grid Generation Using blockMesh

Given computational region is divided into blocks with eight vertices for each block as shown in Figure 7.7. The vertex numbers and block numbers are shown in Figure 7.7. One can see that there are 48 vertices going from 0 to 47 and 17 blocks from 0 to 16 as shown in Figure 7.8. The building block is not considered in the numbering, because that region is not discretized. Each block is identified by 6 faces and 12 edges. First, the coordinates of the vertices are given under "vertices" command. Next, the blocks details are given under "blocks" command. At this time, equal spacing cell with 0.1 H is considered, and hence expansion ratio used in all the three directions to be 1 in the "hex" command under blocks. If expansion ratio 2 is used, the cell spacing is increased from h to 2h, and so on. For the block-1, the width in the x, y, and z directions from Figure 7.8 are (4,3,1), and hence the number of cells in each direction is provided as (40,30,10) in the "hex" command. There are eight blocks at the bottom level by not including the building and nine blocks at the top level. The different boundary faces are identified using the vertex numbers after the block details. The vertex numbers for each block is given in the counterclockwise direction with first the bottom face and then the top face as illustrated in the following section for block-1. More details are given in the user manual.

```
/*--------------------------------*- C++ -*----------------------------------*\
=========                 |
  \\      /  F ield         | OpenFOAM: The Open Source CFD Toolbox
   \\    /   O peration      | Website:  https://openfoam.org
    \\  /    A nd            | Version:  7
     \\/     M anipulation  |
\*---------------------------------------------------------------------------*/
FoamFile
{
    version     2.0;
    format      ascii;
    class       volVectorField;
    object      U;
}
// * * * * * * * * * * * * * * * * * * * * * * * * * * * * * * * * * * * * * //

dimensions      [0 1 -1 0 0 0 0];
internalField   uniform (0 0 0);
boundaryField
{
    inlet
    {
        type            fixedValue;
        value           uniform (1 0 0);
    }
    outlet
    {
        type            zeroGradient;
    }
    Wall
    {
        type            noSlip;
    }
    top
    {
        type            symmetryPlane;
    }
    front
    {
        type            symmetryPlane;
    }
    back
    {
        type            symmetryPlane;
    }
}
// ************************************************************************* //
```

Figure 7.5   Detail file "U" for initial conditions under **0** directory.

The detail of hex command used under blocks command is explained using the first block as follows:

hex (0 4 5 1 16 20 21 17)// vertex numbers of block-1.
(40 30 10) // Number of cells in each direction
simpleGrading (1 1 1) // cell width expansion ratio

```
/*--------------------------------*- C++ -*----------------------------------*\
  =========                 |
  \\      /  F ield          | OpenFOAM: The Open Source CFD Toolbox
   \\    /   O peration      | Website:  https://openfoam.org
    \\  /    A nd            | Version:  7
     \\/     M anipulation   |
\*---------------------------------------------------------------------------*/
FoamFile
{
    version     2.0;
    format      ascii;
    class       volScalarField;
    object      p;
}
// * * * * * * * * * * * * * * * * * * * * * * * * * * * * * * * * * * * * * //
dimensions      [0 2 -2 0 0 0 0];
internalField   uniform 0;
boundaryField
{
    inlet
    {
        type                zeroGradient;
    }
    outlet
    {
        type                fixedValue;
        value               uniform 0;
    }
    top
    {
        type                symmetryPlane;
    }
    Wall
    {
        type                zeroGradient;
    }
    front
    {
        type                symmetryPlane;
    }
    back
    {
        type                symmetryPlane;
    }
}
// ************************************************************************* //
```

Figure 7.6 Detail file "p" for initial conditions under **0** directory.

7.3.6 Detail of "fvSchemes" File

In this file, the finite volume scheme used to approximate the momentum equation both in time and space is selected as shown in Figure 7.9. For the current work, central difference or linear approximation is used. For time, Euler backward method is used. Other methods can be selected depending on the type of problem solved. Other available procedures are available in the user manual Section 6.2.

7.3.7 Detail of "fvSolution" File

In this method, the solution techniques used to solve the simultaneous equations formed out of each momentum equation and pressure equation are specified as shown in

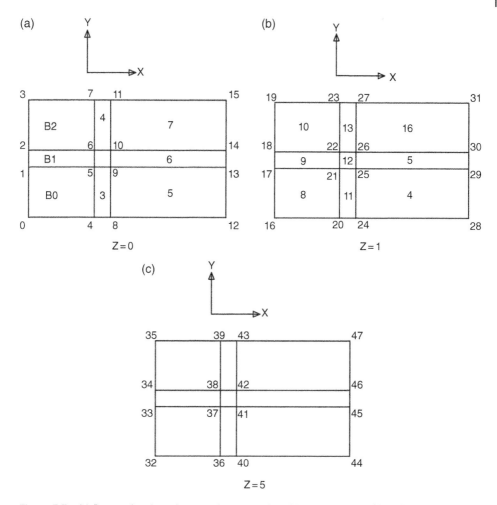

Figure 7.7 (a) Bottom-level vertices number at z = 0 and block numbers. (b) Middle-level vertices number at z = 1 and block numbers and (c) top-level vertices number at z = 5.

Figure 7.10. For the current work, velocities are solved by symmetric Gauss–Siedel and pressure is solved by PCG. The outer iteration used to solve the NS equation is PISO method. The methods available in OpenFOAM are listed and explained in Section 6.3 of the user manual.

7.3.8   Detail of "controlDict" File

In this file, time step and total time to be used and input and output data to be written can be specified as discussed in user manual Section 6.1. The controlDict file is shown in Figure 7.11. The time step deltaT in Figure 7.11 is kept as 0.025. This value comes to a CFL value of about 0.5. This can be calculated as dt = h/Umax = 0.1/2.0 = 0.05. To have a CFL of 0.5, half the value is taken. The details of screen output for a time step will be discussed in the following section.

```
/*--------------------------------*- C++ -*----------------------------------*\|
=========                 |
| \\      /  F ield        | OpenFOAM: The Open Source CFD Toolbox
| \\    /   O peration     | Version:  dev
|  \\  /    A nd           | Web:      www.OpenFOAM.org
|   \\/     M anipulation  |
\*---------------------------------------------------------------------------*/
FoamFile
{
    version     2.0;
    format      ascii;
    class       dictionary;
    object      blockMeshDict;
}
// * * * * * * * * * * * * * * * * * * * * * * * * * * * * * * * * * * * * * //

convertToMeters 1;

vertices
(
    (0 0 0)    //0
    (0 3 0)    //1
    (0 4 0)    //2
    (0 7 0)       //3

    (4 0 0)    //4
    (4 3 0)    //5
    (4 4 0)    //6
    (4 7 0)    //7

    (5 0 0)    //8
    (5 3 0)    //9
    (5 4 0)    //10
    (5 7 0)       //11

    (12 0 0)  //12
    (12 3 0)          //13
    (12 4 0)          //14
    (12 7 0)          //15

    (0 0 1)    //16
    (0 3 1)    //17
    (0 4 1)    //18
    (0 7 1)       //19

    (4 0 1)    //20
    (4 3 1)    //21
    (4 4 1)    //22
    (4 7 1)    //23

    (5 0 1)    //24
    (5 3 1)    //25
    (5 4 1)    //26
    (5 7 1)       //27

    (12 0 1)  //28
    (12 3 1)          //29
    (12 4 1)          //30
    (12 7 1)          //31

    (0 0 5)    //32
    (0 3 5)    //33
    (0 4 5)    //34
    (0 7 5)       //35

    (4 0 5)    //36
    (4 3 5)    //37
    (4 4 5)    //38
    (4 7 5)    //39

    (5 0 5)    //40
    (5 3 5)    //41
    (5 4 5)    //42
    (5 7 5)       //43

    (12 0 5)  //44
    (12 3 5)  //45
    (12 4 5)  //46
    (12 7 5)  //47
);
```

Figure 7.8 Detail file "blockMeshDict" for initial conditions under **system** directory.

```
blocks
(
    hex (0 4 5 1 16 20 21 17) (40 30 10) simpleGrading (1 1 1)
    hex (1 5 6 2 17 21 22 18) (40 10 10) simpleGrading (1 1 1)
    hex (2 6 7 3 18 22 23 19) (40 30 10) simpleGrading (1 1 1)
    hex (4 8 9 5 20 24 25 21) (10 30 10) simpleGrading (1 1 1)
    hex (6 10 11 7 22 26 27 23) (10 30 10) simpleGrading (1 1 1)
    hex (8 12 13 9 24 28 29 25) (70 30 10) simpleGrading (1 1 1)
    hex (9 13 14 10 25 29 30 26) (70 10 10) simpleGrading (1 1 1)
    hex (10 14 15 11 26 30 31 27) (70 30 10) simpleGrading (1 1 1)
    hex (16 20 21 17 32 36 37 33) (40 30 40) simpleGrading (1 1 1)/second level starts
    hex (17 21 22 18 33 37 38 34) (40 10 40) simpleGrading (1 1 1)
    hex (18 22 23 19 34 38 39 35) (40 30 40) simpleGrading (1 1 1)
    hex (20 24 25 21 36 40 41 37) (10 30 40) simpleGrading (1 1 1)
    hex (21 25 26 22 37 41 42 38) (10 10 40) simpleGrading (1 1 1)
    hex (22 26 27 23 38 42 43 39) (10 30 40) simpleGrading (1 1 1)
    hex (24 28 29 25 40 44 45 41) (70 30 40) simpleGrading (1 1 1)
    hex (25 29 30 26 41 45 46 42) (70 10 40) simpleGrading (1 1 1)
    hex (26 30 31 27 42 46 47 43) (70 30 40) simpleGrading (1 1 1)
);

edges
(
);

boundary
(
    Wall
    {
        type wall;
        faces
        (
            (0 4 5 1)
            (1 5 6 2)
            (2 6 7 3)
            (4 8 9 5)
            (6 10 11 7)
            (8 12 13 9)
            (9 13 14 10)
            (10 14 15 11)

            (5 6 22 21)
            (5 9 25 21)
            (9 10 26 25)
            (6 10 26 22)
            (21 22 26 25)
        );
    }

    top
    {
        type symmetryPlane;
        faces
        (
            (32 36 37 33)
            (33 37 38 34)
            (34 38 39 35)
            (36 40 41 37)
            (37 41 42 38)
            (38 42 43 39)
            (40 44 45 41)
            (41 45 46 42)
            (42 46 47 43)
        );
    }
```

Figure 7.8 (Continued)

```
    front
    {
        type symmetryPlane;
        faces
        (
            (0 4 20 16)
            (16 20 36 32)
            (4 8 24 20)
            (20 24 40 36)
            (8 12 28 24)
            (24 28 44 40)
        );
    }
    back
    {
        type symmetryPlane;
        faces
        (
            (3 7 23 19)
            (19 23 39 35)
            (7 11 27 23)
            (23 27 43 39)
            (11 15 31 27)
            (27 31 47 43)
        );
    }
    inlet
    {
        type patch;
        faces
        (
            (0 1 17 16)
            (16 17 33 32)
            (1 2 18 17)
            (17 18 34 33)
            (2 3 19 18)
            (18 19 35 34)
        );
    }
    outlet
    {
        type patch;
        faces
        (
            (12 13 29 28)
            (28 29 45 44)
            (13 14 30 29)
            (29 30 46 45)
            (14 15 31 30)
            (30 31 47 46)
        );
    }
);

mergePatchPairs
(
);

// ****************************************************************** //
```

Figure 7.8 (Continued)

When the case file is run, at each time step, the error for different variables is written either on the screen or in our case stored in the "screen.txt" file. As an example, for the last time step, screen output is shown in the following section. Each time step took about 2 s to run.

```
/*--------------------------------*- C++ -*----------------------------------*\
=========                 |
  \\      /  F ield         | OpenFOAM: The Open Source CFD Toolbox
   \\    /   O peration     | Website:  https://openfoam.org
    \\  /    A nd           | Version:  7
     \\/     M anipulation  |
\*---------------------------------------------------------------------------*/
FoamFile
{
    version     2.0;
    format      ascii;
    class       dictionary;
    location    "system";
    object      fvSchemes;
}
// * * * * * * * * * * * * * * * * * * * * * * * * * * * * * * * * * * * * * //
ddtSchemes
{
    default         Euler;
}
gradSchemes
{
    default         Gauss linear;
    grad(p)         Gauss linear;
}
divSchemes
{
    default         none;
    div(phi,U)      Gauss linear;
}
laplacianSchemes
{
    default         Gauss linear orthogonal;
}
interpolationSchemes
{
    default         linear;
}
snGradSchemes
{
    default         orthogonal;
}

// ************************************************************************* //
```

Figure 7.9   Detail file "fvSchemes" for initial conditions under **system** directory.

```
Time = 10
Courant Number mean: 0.257348 max: 0.491061
smoothSolver:  Solving for Ux, Initial residual = 0.000353437,
Final residual = 2.91519e-06, No Iterations 1
smoothSolver:  Solving for Uy, Initial residual = 0.000631218,
Final residual = 5.15896e-06, No Iterations 1
smoothSolver:  Solving for Uz, Initial residual = 0.000876484,
Final residual = 9.65766e-06, No Iterations 1
DICPCG:  Solving for p, Initial residual = 0.000142511, Final
residual = 6.5553e-06, No Iterations 58
```

```
/*--------------------------------*- C++ -*----------------------------------*\
  =========                 |
  \\      /  F ield         | OpenFOAM: The Open Source CFD Toolbox
   \\    /   O peration     | Website:  https://openfoam.org
    \\  /    A nd           | Version:  7
     \\/     M anipulation  |
\*---------------------------------------------------------------------------*/
FoamFile
{
    version     2.0;
    format      ascii;
    class       dictionary;
    location    "system";
    object      fvSolution;
}
// * * * * * * * * * * * * * * * * * * * * * * * * * * * * * * * * * * * * * * //

solvers
{
    p
    {
        solver          PCG;
        preconditioner  DIC;
        tolerance       1e-06;
        relTol          0.05;
    }

    pFinal
    {
        $p;
        relTol          0;
    }

    U
    {
        solver          smoothSolver;
        smoother        symGaussSeidel;
        tolerance       1e-05;
        relTol          0;
    }
}

PISO
{
    nCorrectors     2;
    nNonOrthogonalCorrectors 0;
    pRefCell        0;
    pRefValue       0;
}
// ************************************************************************* //
```

Figure 7.10   Detail file "fvSolution" for initial conditions under **system** directory.

```
time step continuity errors : sum local = 5.29391e-10, global =
6.22507e-11, cumulative = 7.18296e-07
DICPCG:  Solving for p, Initial residual = 2.12655e-05, Final
residual = 8.6798e-07, No Iterations 96
time step continuity errors : sum local = 7.00758e-11, global =
1.24003e-12, cumulative = 7.18297e-07
ExecutionTime = 863.82 s  ClockTime = 864 s
```

```
/*--------------------------------*- C++ -*----------------------------------*\
=========                 |
  \\      /  F ield        | OpenFOAM: The Open Source CFD Toolbox
   \\    /   O peration    | Website:  https://openfoam.org
    \\  /    A nd          | Version:  7
     \\/     M anipulation |
\*---------------------------------------------------------------------------*/
FoamFile
{
    version     2.0;
    format      ascii;
    class       dictionary;
    location    "system";
    object      controlDict;
}
// * * * * * * * * * * * * * * * * * * * * * * * * * * * * * * * * * * * * * //

application     icoFoam;
startFrom       startTime;
startTime       0;
stopAt          endTime;
endTime         10;
deltaT          0.25e-1;
writeControl    timeStep;
writeInterval   40;
purgeWrite      0;
writeFormat     ascii;
writePrecision  6;
writeCompression off;
timeFormat      general;
timePrecision   6;
runTimeModifiable true;
functions
{
    probes
    {
        type            probes;
        libs            ("libsampling.so");
        writeControl    timeStep;
        writeInterval   1;

        fields
        (
            U p
        );

        probeLocations
        (
            (4.1 3.5 1)
            (4.5 3.5 1)
        );

    }
}
// ************************************************************************* //
```

Figure 7.11 Detail of "controlDict" file for initial conditions under **system** directory.

From the earlier-mentioned write-up, we can monitor how the outer iteration of the NS equation is performing and how the error is varying at each time step. As predicted, we can notice that the max. CFL number is 0.49. In the beginning, the velocities took three iterations to converge to the specified error tolerance and at the end it is only one. The pressure solver took more than 150 iterations to converge in the beginning and later on it started to reduce (58, 96 iterations above). Especially, the second iteration took always more number of iterations to converge. The CPU time in the machine I ran came to be 864s or 14.4m.

7.3.9 Time Variation of Data

The probes function in Figure 7.11 is used to plot the time variation of the pressure at certain points. Here, the two points chosen are (4.1,3.5,1.0) and (4.5, 3.5, 1.0). These two points are on the roof along the midway of the building. The first point is approximately the minimum peak pressure point or close to the windward roof edge. After running the case file, a directory called **post Processing** will be created. Under that you can find the values of U and p under the subdirectory **probes**. The pressure file can be processed for x–y plot as shown in Figure 7.12. From the figure, one can see that the pressure is decreasing as time goes on and the slope has been reducing, and it is not sure that it reached complete steady state. The pressure coefficient C_p from the plot at time 10 is −1.02. Here, C_p is defined as $C_p = 2(p - p_{ref})/(\rho U_{ref}^2)$. Same way if we open the "U" file, one can see the time variation of the velocity at the probing points. It was noticed that there is value for the velocity on the wall, and it is not clear how the OpenFOAM is getting it.

7.3.10 Space Data Retrieval from ParaView

Retrieving data along a line can be done during visualization using ParaView. The detail will be discussed in the following discussions.

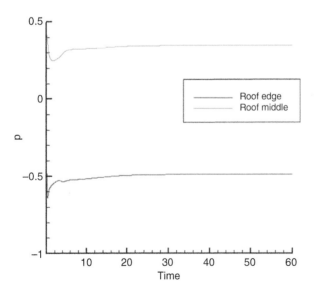

Figure 7.12 Time variation of pressure close to the windward edge of the roof and middle of the roof.

7.4 Visualization Using ParaView

There are several features available in ParaView and as an open-source program, it is very attractive. Many tutorials and user information are available in the web. Here, we will demonstrate how one can plot grid, contours, stream traces, and vectors on a plane. These planes can be moved along the x-, y-, and z-axis. This way one can get good understanding of the flow. The more one spends time to use visualization, they may find many useful features.

7.4.1 Loading the Data from OpenFOAM for Visualization

Run the ParaView program and open the file controlDict under **system** directory. To see the file controlDict, select "All files" option to see it. Then select "OpenFOAM Reader" in the next screen. Then click the eye for controlDict in the Pipeline Browser Box on the left or click "Apply" under "properties" box. In the viewing box, you will see 2D surface of the solution region. In my case, it showed xy plane. At this time, spend time to learn different icon in ParaView and how one can change the 3D view by moving the mouse over the surface. I selected "Y+" icon on the top right side of the ParaView. This changed the view to xz plane. In the top middle, a box showed "suface." One can click and choose other options and see what happened to the view. To store the image as a tiff file, click "file" and then "screen shot" and select where to save under tiff file option. My first image is shown in Figure 7.13. This figure shows pressure value on the outer surface with different colors.

7.4.2 3D view with Grid Axes and Grid Spacing on the Building

To plot the axes as shown in Figure 7.3a using ParaView, use the following steps. Load the data as before and pick wall, back, and outlet, and remove internal mesh in the properties box. Then choose the option of "Data Axes Grid" and apply. Select Y+ icon and rotate for best view of the axes as shown in Figure 7.3a. The grid plot shown in Figure 7.3b uses the similar procedure but keeps only the wall and "select surface with edges" option instead of "surface" to plot.

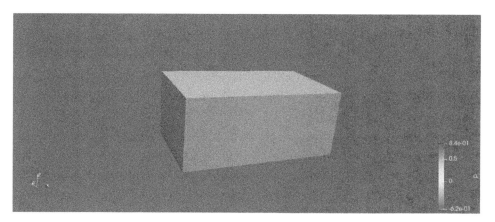

Figure 7.13   3D view of the computational region box.

7.4.3 Contour on xz Slice

To draw the contour on the xz middle plane, click the "slice" icon on the left top corner. Then in the properties box, choose y normal box and then click "apply" (slice-y normal-apply). Then, one can see the pressure contour. If one wants U magnitude contour, choose that in the box instead of "p." The pressure contour and at y = 3.5H is shown in Figure 7.14. By changing the y coordinate of the origin box under properties from (6,3.5,2.5), one can move the slice along the y-axis. Also, by moving the mouse wheel, one can zoom in and out.

7.4.4 Velocity Vector Diagram

To draw the vector diagram on the xz slice, click the "glyph" icon close to slice icon. In the scale array box under "properties," choose "no scale array" and reduce the scale factor as start and then "apply." To zoom, use click zoom box and choose the new box to zoom. To have a better view, adjust the scale factor. The zoom view of the velocity vector with different contour colors are shown in Figure 7.15.

7.4.5 Streamline Plot for xz Slice

Upload the data and select the xz slice at the middle section to cut the building as discussed in the previous sections. Make sure to select the last time step data. Next, click the icon for "stream tracer" next to "glyph" icon. Choose "line parameter" under properties box for point 1 and point 2 to be (5.3,3.5,0) and (5.3,3.5, 5). Also choose number of points to consider along this vertical line to be 70 in the "resolution" box below "line parameter" and then click "apply." The stream line plot is shown in Figure 7.16.

7.4.6 Retrieval or Plotting of Data Along a Line Using ParaView

The variation of velocity and pressure along a line can be done within ParaView. Here, the points considered are on the roof of the cube along the x-axis and at the middle of

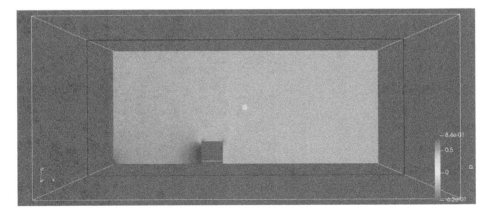

Figure 7.14 Pressure contour at y = 3.5.

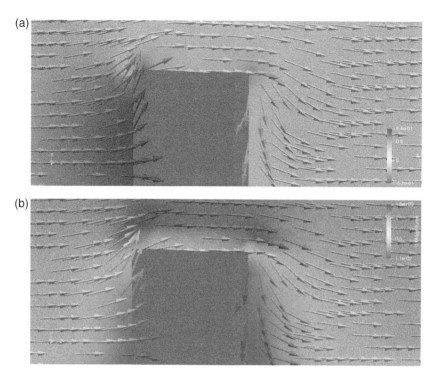

Figure 7.15 (a) Close-up view with pressure as contour color and (b) close-up view with velocity magnitude as contour color.

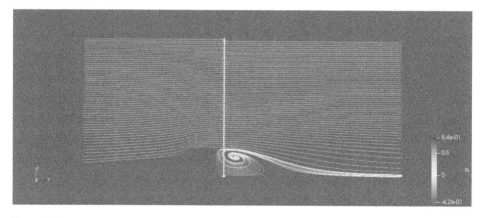

Figure 7.16 Streamline plot for xz slice.

the cube. The coordinates are (4.0,3.5,1.0) and (5.0,3.5, 1.0). The following steps are followed to plot:

1) Open the controlDict file into ParaView and click "apply" in the property box.
2) Click the icon "plot over line" or find the icon under filters (filters-data analysis-plot over line) and then provide the two coordinates under the property box for point1 and point2. Then click apply. You will see a new box opened with u and p plot.

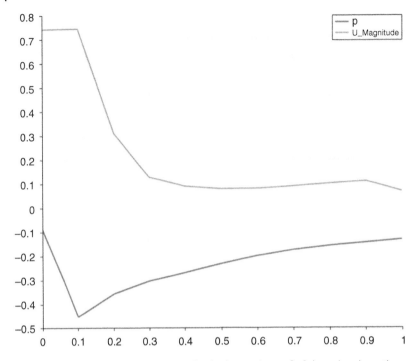

Figure 7.17 Variation of pressure and velocity on the roof of the cube along the x-axis.

3) Click the file and select "save screenshot" to save the tiff file. If one wants to save the data in a txt file, one can select "save data."

One can choose in the property box say 10 points for plot instead 1000 points. The other point to note is to make sure you are choosing the proper time of the file to plot. A sample plot is shown at time = 10 in Figure 7.17. From the plot, one can see that there are velocities on the wall. I am not sure how OpenFOAM works to have velocity on the wall when we specify no-slip BC in Figure 7.5. My understanding is that the all the variables are solved at the cell center. May be these values are coming by projecting the values from the interior.

7.5 Analysis of Flow Over Cube Data for Uniform Flow at the Inlet

The current computer runs are made with 0.1 H grid spacing. The minimum pressure on the top of the roof is −0.49, and the corresponding pressure coefficient is −0.98. Whereas Verma (2020) used a grid spacing of H/160 using snappyHexMesh and the minimum pressure coefficient reported for the same problem with Re = 100 was −1.5. Hence, refined grid reduced the pressure by 50%. These values are at an instant of time. Method has to be devised to calculate min and max from time variation.

7.6   Computation of Turbulent Flow Over a Cube

In this section, we will perform more realistic flow around a cube. Mooneghi et al. (2016) reported a wind tunnel (WT) test on Silsoe building. They compared their WT measurement with field measurements. Here, we will use the same grid used earlier to perform a turbulent flow calculation with proper wind profile at the inlet. As discussed in Chapter 5 for proper modeling, one should consider inflow turbulence. For ease in introducing the concept of large eddy simulation (LES) for turbulence modeling, log-law profile at the inlet and law of the wall boundary condition on the wall will be considered here. We will not consider the inflow turbulence. All these concepts are introduced in the previous chapters, and hence the technical part will not be discussed here. Only the implementation in OpenFOAM will be illustrated. The use of $10 \times 10 \times 10$ cells on the building will not give very good results but to compare with the previous section, we will use the same grid. Mansouri et al. (2021) has some more details in the implementation of LES in OpenFOAM.

The case file used for this work is **buildingLES1**. Here, only part of the case file that has changes will be discussed. The files in different directory for this study are listed in Figure 7.18. The new files added are shown with italic.

7.6.1   Detail of "constant" Directory

In the "transportProperties" file under **constant** directory instead of $\nu = 0.01$ or Re $= 100$, here $\nu = 1.5 \times 10^{-5}$ or Re $= 0.667 \times 10^5$ is used. The "turbulenceProperties" file is listed in Figure 7.19. Here, Smagorinsky LES model is used. The constants connected with this model are provided here.

7.6.2   Detail of "system" Directory

The blockMeshDict file is same as before and hence there is no change. The "controlDict" file is almost same as before except few differences as shown in Figure 7.20. The time step dt = 0.025 is same as before. The simulation is done for 60 time units to see what happens to the flow in time. The pressure p is probed at the same points before. To reduce the number of times, the whole flow is written by increasing the interval to 240 (every 6 time units the

Figure 7.18   Details of **buildingLes1** data or case file.

buildingLes1
    constant
        transportProperties
        turbulenceProperties
    system
        blockMeshDict
        controlDict
        fvSchemes
        fvSolution
    0 [initial values of velocities and p]
        nut
        p
        U

```
\*--------------------------------------------------------------------------*/
FoamFile
{
    version     2.0;
    format      ascii;
    class       dictionary;
    location    "constant";
    object      turbulenceProperties;
}
// * * * * * * * * * * * * * * * * * * * * * * * * * * * * * * * * * * * * * //

simulationType LES;
LES
{
    LESModel         Smagorinsky;
    turbulence       on;
    printCoeffs      on;
    delta            cubeRootVol;
    cubeRootVolCoeffs
    {
        deltaCoeff      1;
    }
    SmagorinskyCoeffs
    {
    Ck          0.094;
    Ce          1.048;
    }
}
// ************************************************************************* //
```

Figure 7.19   List of "turbulenceProperties" file.

files are written). The application is selected as "pisoFoam." That is, piso solver is selected for turbulent flow.

The file "fvSchemes" is listed in Figure 7.21. The finite volume schemes are the same one as before. Only few extra lines are added to take care of the turbulence calculation. The turbulence term in the momentum equations is calculated in the line with "nueff."

The file "fvSolution" is listed in Figure 7.22. Here for pressure solution, instead of DIC-PCG solver used in building, GAMG-Gauss–Seidel solver is used. The solver GAMG is based on algebraic multigrid method. The smoother used is Gauss–Seidel method. In the DIC-PCG solver, the PCG method is used with one form of incomplete Cholesky preconditioner. It is found that for each time step, the GAMG solver took 2 seconds and the DIC-PCG took 3 second. For this particular application, GAMG seems to be an efficient solver. To solve the NS equation instead of "icoFoam" method, "pisoFoam" method is used because the flow is turbulent.

7.6.3   Inflow Details

The nondimensional building height and the wind velocity at the building height are taken as one. The u^* comes to 0.089 for roughness length of 0.01. The logarithmic profile is provided by modifying the "U" file under directory "0" as shown as follows. The important

```
\*---------------------------------------------------------------------------*/
FoamFile
{
    version     2.0;
    format      ascii;
    class       dictionary;
    location    "system";
    object      controlDict;
}
// * * * * * * * * * * * * * * * * * * * * * * * * * * * * * * * * * //
application     pisoFoam;
startFrom       startTime;
startTime       0;
stopAt          endTime;
endTime         60;
deltaT          0.25e-1;
writeControl    timeStep;
writeInterval   240;
purgeWrite      0;
writeFormat     ascii;
writePrecision  6;
writeCompression off;
timeFormat      general;
timePrecision   6;
runTimeModifiable true;
InfoSwitches
{
    writePrecision  6;
    writeJobInfo    0;
    // Allow case-supplied C++ code (#codeStream, codedFixedValue)
    allowSystemOperations   1;
}
functions
{
    probes
    {
        type            probes;
        libs            ("libsampling.so");
        writeControl    timeStep;
        writeInterval   1;

        fields
        (
            p
        );

        probeLocations
        (
            (4.1 3.5 1)
            (4.5 3.5 1)
        );

    }
}
// ************************************************************************* //
```

Figure 7.20   List of "controlDict" file.

```
\*---------------------------------------------------------------------------*/
FoamFile
{
    version     2.0;
    format      ascii;
    class       dictionary;
    object      fvSchemes;
}
// * * * * * * * * * * * * * * * * * * * * * * * * * * * * * * * * * * * //
ddtSchemes
{
    default         Euler;
}
gradSchemes
{
    default         Gauss linear;
    limited         cellLimited Gauss linear 1;
    grad(U)         $limited;
    grad(k)         $limited;
    grad(epsilon)   $limited;
}
divSchemes
{
    default         none;
    div(phi,U)      bounded Gauss linearUpwind limited;
    turbulence      bounded Gauss limitedLinear 1;
    div(phi,k)      $turbulence;
    div(phi,epsilon) $turbulence;
    div((nuEff*dev2(T(grad(U))))) Gauss linear;
}
laplacianSchemes
{
    default         Gauss linear corrected;
}
interpolationSchemes
{
    default         linear;
}
snGradSchemes
{
    default         corrected;
}

wallDist
{
    method meshWave;
}
// ************************************************************************* //
```

Figure 7.21 List of "fvSchemes" file.

point to be noted here in this illustration is that the input is varying in space at the inlet. One can also modify this to include a data varying in time.

```
inlet
{
type codedFixedValue;
redirectType velocity_inlet;
code
#{
```

```
\*---------------------------------------------------------------------------*/
FoamFile
{
    version     2.0;
    format      ascii;
    class       dictionary;
    object      fvSolution;
}
// * * * * * * * * * * * * * * * * * * * * * * * * * * * * * * * * * * //

solvers
{
    p
    {
        solver          GAMG;
        smoother        GaussSeidel;
        tolerance       1e-5;
        relTol          0.1;
    }

    pFinal
    {
        $p;
        tolerance       1e-5;
        relTol          0;
    };

    "(U|k|omega|epsilon)"
    {
        solver          smoothSolver;
        smoother        symGaussSeidel;
        tolerance       1e-6;
        relTol          0;
    }
}

PISO
{
    nCorrectors     2;
    nNonOrthogonalCorrectors 0;
}

// *********************************************************************** //
```

Figure 7.22   List of "fvSolution" file.

```
scalar Ustar=0.089;
scalar k=0.41;
scalar z0=0.01;
fixedValueFvPatchVectorField myPatch(*this);
forAll(this->patch().Cf(),i)
{
myPatch[i]=vector(Ustar/k*(Foam::log((this->patch().Cf()[i].z
())/z0)),0,0);
}
operator==(myPatch);
#};
value $internalField;
}
```

The "nut" file boundary condition is listed in Figure 7.23.

```
\*---------------------------------------------------------------------------*/
FoamFile
{
    version     2.0;
    format      ascii;
    class       volScalarField;
    location    "1";
    object      nut;
}
// * * * * * * * * * * * * * * * * * * * * * * * * * * * * * * * * * * * * * //
dimensions      [0 2 -1 0 0 0 0];
internalField   uniform 0;
boundaryField
{
    inlet
    {
        type            zeroGradient;
    }
    outlet
    {
        type            zeroGradient;
    }
    top
    {
        type            symmetryPlane;
    }
    front
    {
        type            symmetryPlane;
    }
    back
    {
        type            symmetryPlane;
    }
    Wall
    {
        type            nutkWallFunction;
        value           uniform 0;
    }
}
// ***************************************************************************** //
```

Figure 7.23 List of "nut" file.

After running the **buildingLes1** case file using "blockMesh" and "pisoFoam" command, the inlet velocity vector is drawn as shown in Figure 7.24 by choosing inlet under control-Dict and then selecting glyph. Select scale to be U to get proper scaling under glyph. The "pisoFoam" solver is used for turbulent flow using LES. For each time step, the CFD run took about 1s. Total CPU time may be around 40 minutes for $40 \times 60 = 2400$ time steps.

The pressure is probed at the roof edge and middle of the roof as before, and the time variation of the plot for two types of fvSchemes are reported in the following section. The "bounded Gauss linearUpwind limited" used in Figure 7.21 dampens the time variation as shown in Figure 7.25. When the "Gauss linear" scheme used in this case file similar to Figure 7.9 has time variation. To capture the time variation more clearly, longer time of running may be better. From the figure, one can observe that $p_{min} = -0.58$ and $p_{max} = -0.049$ at the roof edge in Figure 7.25b. This gives $p_{ave} = -0.3145$. The upwind limited method gives $p = -0.36$. Compared with previous chapters, these values are much lower

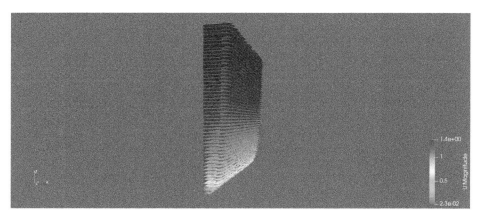

Figure 7.24   View of the inlet velocity profile.

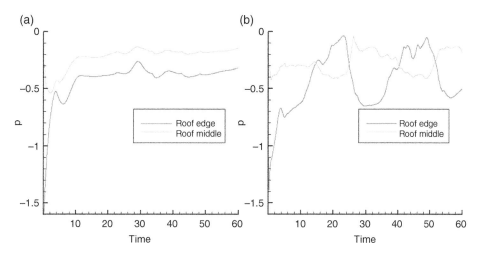

Figure 7.25   Time variation of the pressure on the roof (a) using "bounded Gauss linearUpwind limited" for U and (b) using "Gauss linear" for U.

due to not having enough grid resolution. For pressure coefficient, simply multiply them by 2. To compare this with Re = 100 results, that data has to be run longer time.

7.7   Multilevel Mesh Resolution Using snappyHexMesh Mesh Generator in OpenFOAM

In this book so far, only equal spacing of the grid is used for ease in input data preparation. In rectangular grid system, one can use unequal grid spacing as reported in Selvam (1997) and Verma and Selvam (2020). But for complicated shapes, this type of grid system may not be attractive. For flow around bridges, Selvam et al. (2002) used unstructured grid with a

(a)

(b)

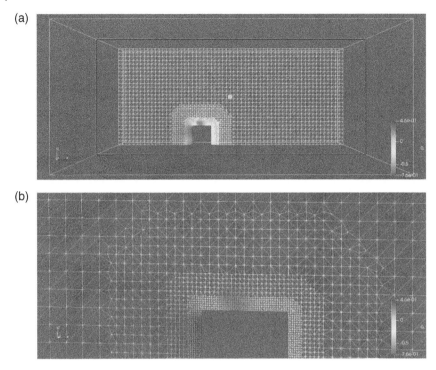

Figure 7.26 Multilevel complex grid generation illustration using three levels. (a) Full computational region and (b) zoomed region around the building.

finite element solver. Due to recent developments in grid generation techniques and numerical methods, one can use finite volume methods with many levels of refinements. The OpenFOAM uses snappyHexMesh tool to generate grid around complex bodies as discussed in Section 4.4 of the user guide (OFUG 2020). To illustrate the grid refinements at several levels, a three-level grid refinement is shown in Figure 7.26 for the same cubic building using the earlier-mentioned grid generator. Here, Level-0 is the starting grid and the next refinement is called Level-1 grid. Hence, there will be three different refinement is done. In this section, we will discuss how a grid using snappyHexMesh can be generated and how CFD solver can be used for this type of grid. For more details, one can also see Verma (2020) class project report available from web.

The performance of multilevel grid like this to gradually varying grid is still not very clear with respect to practical application. At this time, how many grids spacing afterward one has to go for the next level of refinement is not clear. These can be investigated as mini projects once the computational setup is available.

The three-level grid resolution has 202 718 cells and 162 532 grid points. This information can be found out from ParaView under information box. The smallest grid spacing h in the Figure 7.26 is $H/(5 \times 8) = H/40$. As already reported, the domain dimension is $12H \times 7H \times 5H$. For $h = 0.1H$ grid spacing, the number of grid points comes to approximately $120 \times 70 \times 50 = 420\,000$. For $h = H/40$ grid spacing, the total grid points come to 269 million points for equal spacing. The number of grid points using snappyHexMesh is far less than equal spacing (less than 1%). If meaningful results can be obtained, this is an efficient technique.

7.7.1 Procedure to Use snappyHexMesh Mesh Generator

The new files that are added to use snappyHexMesh are in **0, constant,** and **system** directories of the case file as discussed in the following section. The case file for this illustration is called **buildingLES2**.

Step 1: Create an object using CAD or any equivalent software with STL or obj file format. In our case, a unit cube is created as "unitCube.stl" file and the visualization is shown in Figure 7.27. This file is kept in **triSurface** directory under **constant** directory. The origin is kept at west-south-bottom corner of the cube when you stand at the center of the cube. This establishes the origin for the blockMesh grid to be discussed next. It is not clear if the origin of the CAD file and the blockMesh data are connected. The detail of the axes as shown in Figure 7.27 can be shown by turning on the axes box under properties in ParaView. When boundary condition is specified in "U," "p," and "nut" files under **0** directory, the cube is referred as "unitCube." The statement in "U" file is shown as follows:

```
unitCube
    {
        type              noSlip;
    }
```

Step 2: In and around the cube, an equal spacing grid is created using "blockMeshDict" file in **system** directory. This is created with a single block with eight vertices and six faces. The dimension of the region is same as before. This becomes the base grid for flow modeling

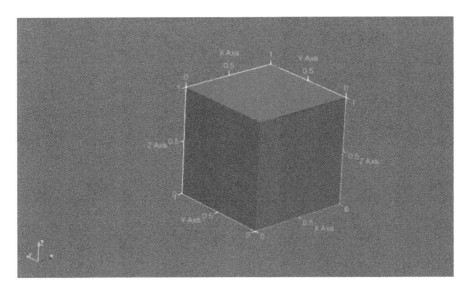

Figure 7.27   Details of the cube with origin.

also. At this time, five cells are provided on each side of the cube as shown in Figure 7.26. The boundary faces at the top, bottom, front, back, inlet, and outlet are identified as we did before. It is not necessary that object (in our case cube) has to go inside exactly with the grid lines. This is explained well in the OpenFOAM user manual example for a car as an object.

Step 3: The "meshQualityDict" file and "surfaceFeaturesDict" file under **system** directory is added to use snappyHexMesh generator. The content is listed in Figures 7.28 and 7.29. From the content, one can see that there is not much for the user need to modify.

Step 4: Most of the activity happens in "snappyHexMesh" file listed in Figure 7.30. The snappyHexMesh also exists in the same **system** directory. Under the "geometry" provide the detail of unitCube mentioned in Step 1. Under "castellatedMeshControls" specify the number of level of refinements, a point in the meshing region and the number of cells beyond which the next fine mesh level start. Here, the corresponding data provided are (3,3), (−1.5, 1.5, 1.5), and 6. Some feature like "addLayers" feature is turned off.

```
\*---------------------------------------------------------------------
*/
FoamFile
{
    version     2.0;
    format      ascii;
    class       dictionary;
    object      meshQualityDict;
}
// * * * * * * * * * * * * * * * * * * * * * * * * * * * * * * * * * * * *
//
#includeEtc "caseDicts/mesh/generation/meshQualityDict.cfg"
//- minFaceWeight (0 -> 0.5)
//minFaceWeight 0.02;
// ******************************************************************
//
```

Figure 7.28 Listing of "meshQualityDict" file.

```
FoamFile
{
    version     2.0;
    format      ascii;
    class       dictionary;
    object      surfaceFeaturesDict;
}
// * * * * * * * * * * * * * * * * * * * * * * * * * * * * * * * * * * * *
//
surfaces ("unitCube.stl");
#includeEtc "caseDicts/surface/surfaceFeaturesDict.cfg"
```

Figure 7.29 Listing of "surafceFeaturesDict" file.

```
FoamFile
{
    version     2.0;
    format      ascii;
    class       dictionary;
    object      snappyHexMeshDict;
}
// * * * * * * * * * * * * * * * * * * * * * * * * * * * * * * * * *
//
#includeEtc "caseDicts/mesh/generation/snappyHexMeshDict.cfg"
castellatedMesh on;
snap            on;
addLayers       off;

geometry
{
    unitCube
    {
        type triSurfaceMesh;
        file "unitCube.stl";
    }
};

castellatedMeshControls
{
    features
    (
      { file "unitCube.eMesh"; level 1; }
    );
    refinementSurfaces
    {
        unitCube
        {
            level (3 3);
        }
    }
    refinementRegions
    {

    }
    locationInMesh (-1.5 1.5 1.5); // Offset from (0 0 0) to avoid
                                   // coinciding with face or edge
    nCellsBetweenLevels 6;
}

snapControls
{
  explicitFeatureSnap     true;
  //implicitFeatureSnap     true;
}
```

Figure 7.30   Listing of "snappyHexMesh" file.

```
addLayersControls
{
    layers
    {
        "unitCube.*"
        {
            nSurfaceLayers 6;
        }
        bottom
        {
            nSurfaceLayers 6;
        }
    }
    relativeSizes        true; // false, usually with firstLayerThickness
    expansionRatio       1.1;
    finalLayerThickness 0.5;
    minThickness         1e-3;
    // firstLayerThickness 0.01;
//   maxThicknessToMedialRatio 0.6;
}
meshQualityControls
{
    minTetQuality -1e+30;
}

/*writeFlags
(
//     scalarLevels
    layerSets
    layerFields
); */
mergeTolerance 1e-6;
// ********************************************************************* //
```

Figure 7.30 (Continued)

7.7.2 Running the Case File buildingLES2

Once the case files are set properly, one can run by executing the following commands under the **buildingLES2** directory:

$blockMesh	[To generate the coarse mesh using blockMesh]
$surfaceFeatures	[This has to be done before executing the snappyHexMesh command]
$snappyHexMesh -overwrite	[To create the multilevel grid]
$pisoFoam	[to run the CFD solver]

In the execution of "snappyHexMesh" step, there is an extension command "overwrite" is there. If that is not added, OpenFOAM created two extra directories (0.0125 and 0.025) as a multiple of time step provided and they are not compatible with other time directories written during "pisoFoam" command and that created visualization issues. The other alternative is to delete those files before one use the "pisoFoam" command.

7.7.3 Selection of Time Step and Total Computational Time

For three-level mesh refinement (L3Mesh), the smallest grid spacing is $h = (H/5)(1/8) = H/40$ and for five-level mesh refinement (L5Mesh), it is $h = (H/5)(1/32) = H/160$. The time step to have a CFL condition of 0.5 is calculated as $h/(2U_{max})$. We will assume U_{max} to be 2 for an inlet velocity of 1 at the cube height and the dt comes to be $1/80 = 0.0125$ and $1/320 = 3.125e\text{-}3$ for CFL = 0.5.

For L3Mesh, the time step 0.0125 worked well but for L5Mesh the time step 3.125e-3 diverged at the second time step during running the job. The reason being, the max. CFL reported to be more than 3 for L5Mesh run. CFL comes to more than 3. Then a time step of 0.001 worked out well.

Computer time for each time step comes to around 0.3 and 3 seconds for H/40 and H/160 grid, respectively. The runs were made for a total time of 50 time units. The data is written for every 10 time units for visualization. The computer time for H/10 grid for full domain, H/40, and H/160 grid with snappy mesh are 36 minutes, 22 minutes, and 61 hours, respectively.

Comparison of Pressure Using H/10 Grid and H/5 grid with Three- and Five-Level snappyHexMesh

For observing the pressure variation in time on the cube, probing coordinates along the center line of the cube (y = 0.5) are recorded at 17 points as shown as follows:

//wind ward wall
Probe 0 (0 0.5 0),   # Probe 1 (0 0.5 0.2) # Probe 2 (0 0.5 0.4) # Probe 3 (0 0.5 0.6)
Probe 4 (0 0.5 0.8)
//roof
Probe 5 (0 0.5 1)   # Probe 6 (0.1 0.5 1) # Probe 7 (0.2 0.5 1) # Probe 8 (0.4 0.5 1)
Probe 9 (0.6 0.5 1)   # Probe 10 (0.8 0.5 1)
//end of roof and leeward wall starts
Probe 11 (1 0.5 1)   # Probe 12 (1 0.5 0.8) # Probe 13 (1 0.5 0.6) # Probe 14 (1 0.5 0.4)
Probe 15 (1 0.5 0.2) # Probe 16 (1 0.5 0)

The pressure contour on the xz slice is shown for both Level 3 (L3Mesh) and Level 5 (L5Mesh) refinements in Figure 7.31. Somewhat bigger vortices are noticed in the L3Mesh comparing to L5Mesh. These plots were created at different times, but the time variation plot in the following section gives much better perspective. The vortex size difference could be due to grid resolution issues. More refined study may shed light on this issue. Here, the main focus is the illustration of using multilevel mesh refinements.

The time variation of pressure at probe points 5 to 8 are plotted in Figure 7.32 for L3Mesh and L5Mesh. These points are on the roof as listed in the preceding section. In Figure 7.32a for L3Mesh, the minimum pressure after 10 seconds seems to be close to −0.6 and it occurs at P7 point. Whereas in Figure 7.32b for L5Mesh, the minimum pressure seems to be little above −0.55 and it occurs at P5 point or close to the windward edge of the roof. The L5Mesh has longer wavelength than L3Mesh pressure variation. The peak pressure is similar to the H/10 grid for uniform inflow.

(a)

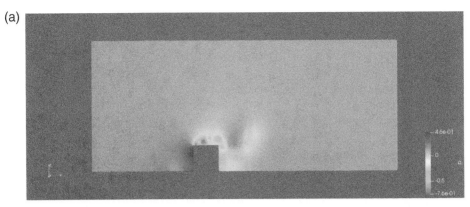

(b)

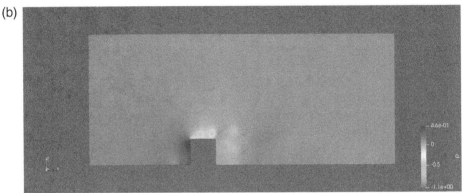

Figure 7.31 Comparison of pressure contour for (a) Level 3 mesh and (b) Level 5 mesh.

7.8 Challenges in Using OpenFOAM

Three example problems are used to show how one can use OpenFOAM for wind engineering applications. The challenges faced in using OpenFOAM as well as using ParaView as visualization are listed as follows:

1) Unless one learns the inner working of the code or learning how to add our own code to do certain work, it seems like a black box. Each research group is using their own version of case file for their application, and the detail of modifications and implementations are sketchy. There are several groups teaching how to use OpenFOAM with a cost and we do not know how much value out of those courses.
2) ParaView as a visualization tool is not as user-friendly as Tecplot. Attempt is made to implement certain functions. We found the velocity vector diagram is not showing clearly when you zoom certain region as inflow around the building. By spending more time, one can learn more ways to eliminate these challenges.

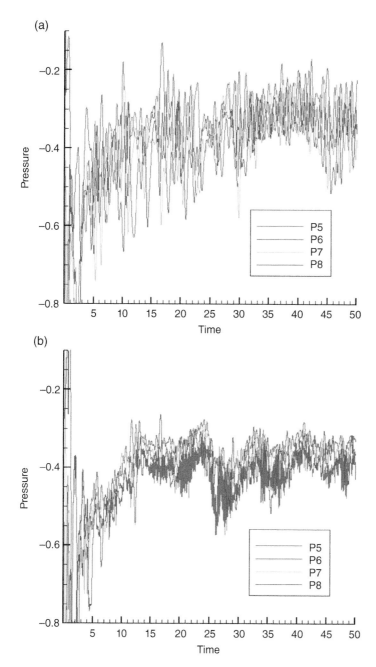

Figure 7.32 Time variation of pressure at probing points 5 to 9 from (a) Level 3 mesh and (b) Level 5 mesh.

7.9 Summary and Conclusions

In this chapter, attempt is made to introduce OpenFOAM for wind engineering. Ways to implement OpenFOAM and ParaView are provided. A simple flow over cubic building is illustrated with visualization, and procedure to run the case file is explained. Further incorporation of log-law profile as inlet, law of the wall condition, and detail of invoking LES is explained in the corresponding case files. Other relevant material for learning available in the web is provided.

The tutorial **WindAroundBuildings** from OpenFOAM illustrates the following important features:

1) Generating snappyHexMesh with many layers of refinement.
2) Use of CAD diagram for complex building shape.
3) Uses k-ε turbulence model.
4) The case file used "simpleFoam" steady solver. Here, one needs to provide underrelaxation factors for u and p. Here, upwind scheme is used for gradient.

Because of steady state solver, the amount of computing time is reduced extensively. For many practical problems in the industry, the k-ε turbulence model is the work horse. Once the LES is well established and the computer cost becomes much cheaper, this may become the common turbulence model for industry.

At this time, incorporation of inflow turbulence is not provided. The paper by Melaku and Bitsuamlak (2021) provides an inflow turbulence generator, corresponding case file, and method to run using OpenFOAM. This is available in the GitHub website: https://github.com/GBitsuamlak/DFSR. One can learn from that how to model using inflow turbulence at the inlet. There is extensive information on OpenFOAM available from the web, and one can expand their knowledge of using OpenFOAM.

7.10 Chapter Outcome

The use of OpenFOAM for wind engineering application is illustrated with case files for three different categories:

1) Laminar flow around a building
2) LES flow around a building
3) Multilevel mesh using LES

Open-source visualization program ParaView is also illustrated using OpenFOAM output files.

Problems

1 Run the case file **building** and get the case file with computer runs.

2 Probe different points and plot in time.

3   Implement all the visualization performed in this chapter.

4   Plot xy stream line and vector plot at different levels say z = 0.5, 0.75, and 1.0

5   Run the case file buildingLes1 and plot the time vs. plot for total time of 90 units.

6   Create a different grid spacing by modifying blockMeshDict file and run the uniform flow case with different Re.

7   Construct a case file for flow around 2D rectangular or circular cylinder. Compare the drag and lift coefficients and Strouhal number St (fD/U) with other reported values as discussed in Chapter 5 for Re = 100. Use the OpenFOAM UserGuide to learn and to create 2D mesh.

References

Liu, X. and Zhang, J. (2019). *Computational Fluid Dynamics: Applications in Water, Wastewater, and Storm Water Treatment*. ASCE Publication.

Mansouri, Z., Verma, S., and Selvam, R.P. (2021). Teaching modeling turbulent flow around building using LES turbulence method and open-source software OpenFOAM. Paper presented at *2021 ASEE Midwest Section Conference, Virtual*, November. 10.18260/1-2-1125.1128.1153-38326. https://peer.asee.org/teaching-modeling-turbulent-flow-around-building-using-les-turbulence-method-and-open-source-software-openfoam (accessed 10 March 2022).

Melaku, A.F. and Bitsuamlak, G.T. (2021). A divergence-free inflow turbulence generator using spectral representation method for large-eddy simulation of ABL flows. *Journal of Wind Engineering and Industrial Aerodynamics* 212: 104580.

Mooneghi, M.A., Irwin, P., and Chowdhury, A.G. (2016). Partial turbulence simulation method for predicting peak wind loads on small structures and building appurtenances. *Journal of Wind Engineering and Industrial Aerodynamics* 157: 47–62.

Moukalled, F., Mangani, L., and Darwish, M. (2016). *The Finite Volume Method in Computational Fluid Dynamics: An Advanced Introduction with OpenFOAM® and Matlab*. Springer.

OFBT (2019). OpenFOAM basic training. 5e. https://www.cfd.at/sites/default/files/tutorialsV7/OFTutorialSeries.pdf (accessed 10 March 2022).

OFUG (2020). OpenFOAM: The opensource CFD tool box, User Guide, version v2012. https://www.openfoam.com/documentation/user-guide (accessed 10 March 2022).

Selvam, R.P. (1997). Computation of pressures on Texas Tech Building using large Eddy simulation. *Journal of Wind Engineering and Industrial Aerodynamics* 67 & 68: 647–657.

Selvam, R.P., Govindaswamy, S., and Bosch, H. (2002). Aeroelastic analysis of bridges using FEM and moving grids. *Wind & Structures* 5: 257–266.

Verma, S. (2020). OpenFOAM for Wind Engineering. Report prepared for a class under Dr. R.P. Selvam at the University of Arkansas.

Verma, S. and Selvam, R.P. (2020). CFD to VorTECH pressure-field comparison and roughness effect on flow. *Journal of Structural Engineering* 146 (9): 04020187-1–04020187-12.

Verma, S., Mansouri, Z., and Selvam, R.P. (2021). Incorporating two weeks open source software lab module in CFD and fluids courses. Paper presented at *2021 ASEE Midwest Section Conference, Virtual*, November. 10.18260/1-2-1125.1128.1153-38325https://peer.asee. org/incorporating-two-weeks-open-source-software-lab-module-in-cfd-and-fluids-courses (accessed 10 March 2022).

Appendix A

Tecplot for Visualization

Tecplot can be used for x–y graphs, two-dimensional (2D) visualization (contour, vector plot, and zooming), three-dimensional (3D) visualization and animation. The grid can be structured or unstructured. In this appendix, several methods for visualization are documented.

For large data visualization: When directly loading the asc plt files, tecplot converts the file to binary for visualization. For large data files such as 3D data, it takes more computer time to load the data. If it is going to be used several times for visualization, it is better to convert to binary using the program **preplot.** The command to convert:

```
>preplot file.plt file_b.plt
```

For extracting data at certain points in the domain, one can use **probing** option. The icon probing can be selected for this or select **probe at** by going under data. The probing can be done with the IJ index option or x and y coordinate option. Also, one can click the grid point and get the detail. The contours can be drawn with flooding or lines or both. Once can also draw streamlines using **stream trace** option. This will help to find the maximum vorticity strength location.

Extracting data on a line using tecplot: use the option **Data-extract-points from polyline**. When you choose say two points and then double click to get out. It will ask where to store. Select files and save as graph.plt. This you can use it as regular tecplot for x–y.

Tecplot options for further data manipulation: At this time, only u, v, and p can be visualized. The vorticity can be calculated using TECPLOT. This can be done using the option **analyse-calculate variables-z-vorticity.** *Before this select under **analyse-field variables-velocity**.*

Computing new variable under tecplot: Let us say we have Sx and Sy as the derivative of stream function S from FDM or FEM analysis.

The relationship for velocities u & v with S are u = dS/dy and v = −dS/dx.

Under tecplot, select specify equations to create new variable v using the following path:

data-alter-specify equations

In the box type, the following and choose compute:

```
{v}=    -{Sx}
```

Computational Fluid Dynamics for Wind Engineering, First Edition. R. Panneer Selvam.
© 2022 John Wiley & Sons Ltd. Published 2022 by John Wiley & Sons Ltd.

Compute

For vector plo, choose U = Sy & V = V and plot velocity vector.

Animation of Data Using Tecplot

Hints for Movie Making: To make movies one need to store different time data as mv1.plt, mv2.plt etc. Some tips to be considered:

1) One should never go for more than 120 movie files. So, make sure you choose the proper interval for IMOVIE.
2) When running the CFD programs, make sure you have the move file names in a directory like **char.txt** file in the same directory and read them.
3) Once the movie files are written at several time step, then the animation is made using tecplot and a macro file called uabm-1.mcr. Before running the mcr file, uabm-1.mcr file has to be edited for proper path, number of movie files, name of the layout file, and name of the movie file.
4) The movie files can be made either avi format or rm format. To run the rm format files, one may need the framer.exe file. Using this format, one can visualize the movie file with many options. Generally, one makes the movie file with avi option.

```
Sample uabm-1.mcr file:
#!MC 800
$!VarSet |NumFiles| = 40
$!EXPORTSETUP EXPORTFORMAT = AVI
$!ExportSetup ExportFName ='D:\rps\f16\research\cwe\genex
\uab-serial-7-28-16\uab-serial-movie\uabm-1.avi'
$!Loop |NumFiles|
$!OpenLayout 'D:\rps\f16\research\cwe\genex\uab-serial-7-28-16
\uab-serial-movie\uabm-1.lay'
  AltPlotFNames = 'D:\rps\f16\research\cwe\genex\uab-serial-
7-28-16\uab-serial-movie\mv|LOOP|.plt'
$! IF |LOOP|==1
  $!EXPORT
      APPEND =NO
$!ENDIF
$! IF |LOOP| !=1
  $!EXPORT
     APPEND =yes
$!ENDIF
$!Endloop
$!quit
```

Creating a uabm-1.lay file using tecplot: The layout file is created using tecplot with one of the movie files. This sets the screen features with whatever one likes to see with all the movie files. Contour range, velocity vector length, and the range of length in x and y directions have to be set using a sample movie file and then they are stored as uabm-1.lay.

Once everything is ready one can simply run the uabm-1.mcr file using tecplot. This will create the movie file in the same directory.

In **tecplot** visualization – to plot more than one screen together use the following action:

1) Under frame-select **create new frame**. Then in an open space using the mouse create a window.
2) Then draw anything you like there. This will get more than one similar figure in the same figure.
3) To keep them in an order, select **frame-tile frame**. The select one later one or any style you like.

Appendix B

Random Process for Wind Engineering

A recorded wind data U(t) in time is called stationary or nonstationary process. The straight wind in the atmospheric boundary layer is assumed to be stationary process. The tornado and thunderstorm wind are considered to be nonstationary process. In a stationary process, the statistical properties are independent of time and space. The nonstationary process is not independent of time and space. The reason for it, there is a sudden wind velocity change in time for nonstraight wind flows. For further understanding, one can refer to other wind engineering or random process books. Here we will discuss mainly stationary process.

Mean (U_{ave}), Variance (σ_u^2) and Standard Deviation (σ_u)

Given a velocity data U(t) for **np** time step with constant time step of dt, one can calculate **mean U_{ave}** and **variance** σ_u^2. They are defined as follows:

$$U_{ave} = \sum U(t)/np, \text{ mean-square} = U_{rms}^2 = \sum U^2(t)/np,$$

where rms means root mean square.

The measure of how much U(t) differs from the average is given by variance σ_u^2 and is defined as

$$\sigma_u^2 = \sum [U(t) - U_{ave}]^2/np = U_{rms}^2 - U_{ave}^2.$$

Here σ_u is the **standard deviation**.

Turbulence Intensity and Reynolds Stresses: The standard deviation of velocities in the x, y, and z directions are represented by σ_u, σ_v, and σ_w. They are related to u^* as follows:

$$\sigma_u, = Au*, \sigma_v = 0.75\,\sigma_u, \text{ and } \sigma_w = 0.5\,\sigma_u.$$

where A = 2.5 if z_0 = 0.05 m and A = 1.8 for z_0 = 0.3 m as reported in Dyrbye and Hansen (1999). The turbulence intensity I_u is defined as a nondimensional number:

$$I_u = \sigma_u/U_{ave}$$

This value varies from 0.2 to 0.1. Close to the ground I value is high. For a WT test data from FIU-WOW, the values are

Device name: Cobra 239
Device type: Four-hole Cobra Probe
Device ID: 239

First sample date: 19-Aug-19
First sample time: 09:50:50.922

Sampling time (s): 301.466
Number of samples: 753 664
Number of good samples: 753 664 (100.0%)
Data output rate (Hz): 2 500.0
Output block size: 4 096
Mean temperature (°C): 30.6
Barometric pressure (Pa): 101 826.4
Mean reference pressure: 0.0

Mean flow speed, pitch angle, yaw angle, and Pstatic (m/s, ° and Pa):

20.7	0.2	4.4	76.8

Mean U, V, and W (m/s):

20.3	1.71	0.102

Minimum and maximum velocities (m/s):

6.35	36.9

Turbulence intensities – overall, Iuu, Ivv, Iww (%):

13.6	15.8	11.8	12.9

Reynolds normal stresses – Ruu, Rvv, Rww (Pa):

−12.5	−6.95	−8.30

Reynolds shear stresses – Ruv, Ruw, Rvw (Pa):

−3.84	−0.394	−0.4798.30

In the preceding data, the Reynolds stresses are given in actual units. Usually these values are represented nondimensionally by dividing by U_{ave}^2. As an example:

$$Ruu = \rho \sigma_u^2 = 12.5 \, (Pa),$$

then $Ruu/\rho \, U_{ave}^2 = 12.5/[1.225(20.3)^2] = 0.0248 = I^2u$.

Hence $Iu = 0.157$, which is close to the reported value of 0.136.

Correlation: Correlation is a measure of the relationship or dependence between two random processes. Let us say we have two velocity records $u_1(t)$ and $u_2(t)$. Here u_1 and u_2 represent velocity data with mean zero. If the two values are multiplied at each time and then take the average over all the points as follows:

$$cx = \left(\sum u_1 u_2\right)/np$$

If the cx is high, then we say they have high correlation. If there is not much correlation, then the cx value approach zero.

Autocorrelation and cross-correlation functions: If the same $u(t)$ is taken with a shift in time say $u(t + \tau)$ and then calculate the correlation cx, and this is called autocorrelation $R(\tau)$.

$$R(\tau) = \left(\sum u(t)u(t + \tau)\right)/np$$

Hence when $\tau = 0$, then $R(0) = \sigma_u^2$.

Usually the autocorrelation function is normalized by dividing the σ_u^2, this can be represented as $\rho_u(\tau) = R(\tau)/\sigma_u^2$, and this is plotted in Figure B.1.

Same way instead of time, if space is considered for correlation between two points, then it is called cross correlation function. The normalized cross-correlation between two points at a distance of rx in the x-direction for u velocity is $\rho_u(rx)$.

Integral time and length scale: The integral time scale is defined as

$$\tau_u = \int \rho_u(\tau)d\tau$$

This is also shown in Figure B.1. The integral time scale represents the time up to which perturbations of u can be correlated.

The integral length scale is defined as

$$L_u^x = \int \rho_u(rx)drx$$

The integral length scales are the average size of the vortices in the wind. From Taylor's hypothesis of frozen turbulence, the time and length scale can be related as $L_u^x = U_{ave} \tau_u$. There are totally nine integral length scales depending upon the velocity and direction

Figure B.1 Autocorrelation function and integral time scale.

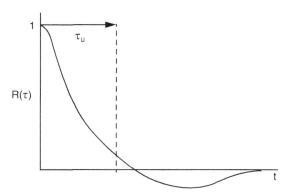

considered. The longitudinal integral length scale L_u^x may vary from 10 to 240 m. The other length scales can be represented in terms of L_u^x as $L_u^y = 0.3\ L_u^x$ and $L_u^z = 0.2L_u^x$ as reported in Dyrbye and Hansen (1999). Atmospheric measurements of length scales are reported in Emes et al. (2018).

Mooneghi et al. (2016) reported that the integral length scale in the field for TTU to be 35 m and in WT 0.43 m.

References

Dyrbye, C. and Hansen, S.O. (1999). *Wind Loads on Structures*. New York: John Wiley & Sons.

Emes, M.J., Jafari, A., and Arjomandi, M. (2018). Estimating the turbulence length scales from cross-correlation measurements in the atmospheric surface layer, *21st Australasian Fluid Mechanics Conference*, Adelaide, Australia (10–13 December).

Mooneghi, M.A., Irwin, P., and Chowdhury, A.G. (2016). Partial turbulence simulation method for predicting peak wind loads on small structures and building appurtenances. *Journal of Wind Engineering and Industrial Aerodynamics* 157: 47–62.

Appendix C

Direct Solution of Ax = b by A⁻¹

For small size A matrix, one can calculate A^{-1} using excel. Then the solution of a simultaneous equations $Ax = b$ can be calculated by $x = A^{-1}b$ matrix operation. This is not an efficient procedure. For efficient procedure, one should use Gaussian elimination or Cholesky decomposition.

Step 1 – Create matrices A and b: Enter the A matrix and the b vector as shown later.

	A	B	C	D	E	F	G
1							
2	A–matrix					b–RHS	
3	4	−1	0	−1		1	
4	−1	4	−1	0		1	
5	0	−1	4	−1		2	
6	−1	0	−1	4		2	
7							

Step 2 – Calculate A⁻¹: Select cells where A^{-1} has to be calculated. Then use the function MINVERSE to invert it. After entering the formula as shown later, use **ctrl + shift + enter** to fill the numbers.

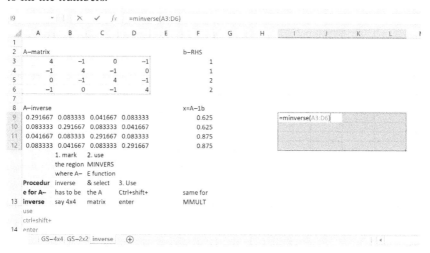

Computational Fluid Dynamics for Wind Engineering, First Edition. R. Panneer Selvam.
© 2022 John Wiley & Sons Ltd. Published 2022 by John Wiley & Sons Ltd.

Step 3 – Perform the operation x = A⁻¹b using MMULT: As in Step 2, select the space where you want x matrix. Then use MMULT formula and select A^{-1} and b matrices. After entering the formula as shown later, use **ctrl + shift + enter** to fill the numbers. The answer is shown in later x.

Calculation of determinant: One can calculate determinant using MDETERM function.

Index

Computational Fluid Dynamics for Wind Engineering, First Edition. R. Panneer Selvam.
© 2022 John Wiley & Sons Ltd. Published 2022 by John Wiley & Sons Ltd.

w1dcn.f-Wave1D-Crank-Nicolson   61
yif2.f-3D turbulent flow   108

r

Random Fourier method   122
Random process   217
   auto correlation   219
   cross-correlation   219
   integral length scale   219
   integral time scale   219
   turbulence intensity   34, 217
Recycling method   121
Reduced velocity   83
Relaxation factor (RF)   47
Resonance   83
Reynolds number (Re)   15
Reynolds stresses   34, 83
Roughness length   78

s

Sliding mesh method   162
snappyHexMesh   175
SIMPLE   178
Staggered grid system   68
Storage methods   69
Stream function   12, 42
Strouhal number   83
Structured grid system   68
Sub-iteration   67
Successive over relaxation (SOR)   46
Synthetic turbulence method   121

t

Tecplot   213
   animation   214

Tent function   23
Topographic effect   78
Tornado-like vortex   3
Tornado-structure interaction   162
   analytical vortex model   162
   terrain effect on tornado path   163
   tornado-dome interaction   163
   tornado models   162
   vortex generation chamber   166
Turbulence intensity   34, 217
Turbulence modeling   32
   direct numerical simulation (DNS)   32
   k-ε turbulence model   34
   large eddy simulation (LES)   32, 35
   RANS equation   32
   Zero-equation model   34

u

Unsteady problem   55
   explicit procedure   58
   implicit procedure   59

v

Velocity pressure   79
Velocity spectrum   20
Visualization   175
Vorticity   12

w

Wave equation   55
Wind-bridge interaction   1
Wind spectrum   28, 134
Wind tunnel (WT)   4
Windows operating system   176
Wind turbine   160